Searching for the Just C

Cities are many things. Among their least appealing aspects, cities are frequently characterized by concentrations of inequality, insecurity, and exploitation. Cities have also long represented promises of opportunity and liberation. Public decision-making in contemporary cities is full of conflict, and principles of justice are rarely the explicit basis for the resolution of disputes. If today's cities are full of injustices and unrealized promises, how would a Just City function? Is a Just City merely a utopia, or does it have practical relevance in a global and dynamic world? Is it the best formulation of the most desirable goal for urban development? This book engages with the growing debate around these questions.

The notion of the Just City emerges from philosophical discussions about what justice is, combined with the intellectual history of utopias and ideal cities. The contributors to this volume, including Susan Fainstein, David Harvey, Peter Marcuse, and Margit Mayer, articulate a conception of the Just City and then examine it from differing angles, ranging from Marxist thought to communicative theory and neoliberal urbanism. The arguments both develop the concept of a Just City and question it, as well as suggesting alternatives for future expansion. Explorations of the concept in practice include case studies primarily from U.S. cities, but also from Europe, the Middle East, and Latin America.

The authors find common ground in the conviction that more far-reaching changes are required in the development of our cities than planners and other urban professionals contemplate. They find that a forthright call for justice in all aspects of city life, putting the question of what a Just City should be on the day-to-day agenda of urban reform, can be a practical approach to solving concrete questions of urban policy, from what to do with a wholesale food market in the Bronx to avoiding gentrification around mega-projects. This synthesis is provocative and timely in a globalized world of ever-changing urban spaces as the contributing authors bridge the gap between theoretical conceptualizations of urban justice and the reality of planning and building cities. The notion of the Just City is an empowering framework for contemporary urban actors to improve the quality of urban life and *Searching for the Just City* is a seminal read for practitioners, professionals, students, researchers, and anyone interested in what urban futures should aim to achieve.

Peter Marcuse, a lawyer and urban planner, is Professor Emeritus of Urban Planning at Columbia University in New York City. He has been involved with urban policy for many years. He was Professor of Urban Planning, first at the University of California at Los Angeles from 1975 to 1978, since then at Columbia University. He has long-standing interests in globalization, comparative housing and planning policies. He is currently involved in, and has written on, the impact of September 11 on New York City, of Katrina on New Orleans, and on globalization, focusing on its impact on social justice.

James Connolly is a doctoral student in Urban Planning at Columbia University. His research currently focuses on urban institutional change at the intersection of environmentalism, community development, and urban planning. Recent publications focus on policy networks in U.S. cities and on spatial patterns of advanced service firms.

Johannes Novy is currently finishing his Ph.D. in Urban Planning at Columbia University's Graduate School for Architecture, Planning and Preservation. He is also a fellow at the Center of Metropolitan Studies (CMS) in Berlin. Novy's research interests include planning history and theory as well as urban development and tourism in North America and Europe.

Ingrid Olivo is a Ph.D. student in Urban Planning at Columbia University. Her research focuses on the role of cultural heritage in post-disaster development planning.

Cuz Potter is a doctoral student in the Urban Planning Department at Columbia University and holds Master's degrees from Columbia in international affairs and urban planning. His research currently centers on utopian thought, social justice, and the relationship between logistics and economic development.

Justin Steil is a joint Ph.D./J.D. student in Urban Planning and in Law at Columbia University. His research focuses on the exercise of power through control over space, especially through the relation between housing, land use and immigration.

'Questioning Cities'

Edited by Gary Bridge, *University of Bristol*, UK and
Sophie Watson, *The Open University*, UK

The 'Questioning Cities' series brings together an unusual mix of urban scholars under the title. Rather than taking a broadly economic approach, planning approach or more socio-cultural approach, it aims to include titles from a multi-disciplinary field of those interested in critical urban analysis. The series thus includes authors who draw on contemporary social, urban and critical theory to explore different aspects of the city. It is not therefore a series made up of books which are largely case studies of different cities and predominantly descriptive. It seeks instead to extend current debates, through in most cases, excellent empirical work, and to develop sophisticated understandings of the city from a number of disciplines including geography, sociology, politics, planning, cultural studies, philosophy and literature. The series also aims to be thoroughly international where possible, to be innovative, to surprise, and to challenge received wisdom in urban studies. Overall it will encourage a multi-disciplinary and international dialogue always bearing in mind that simple description or empirical observation which is not located within a broader theoretical framework would not – for this series at least – be enough.

Published:

Global Metropolitan
John Rennie Short

Reason in the City of Difference
Gary Bridge

In the Nature of Cities
Urban political ecology and the politics of urban metabolism
Erik Swyngedouw, Maria Kaika and Nik Heynen

Ordinary Cities
Between modernity and development
Jennifer Robinson

Urban Space and Cityscapes
Christoph Lindner

City Publics
The (dis)enchantments of urban encounters
Sophie Watson

Small Cities
Urban experience beyond the metropolis
David Bell and Mark Jayne

Cities and Race
America's new black ghetto
David Wilson

Cities in Globalization
Practices, policies and theories
Peter J. Taylor, Ben Derudder, Piet Saey and Frank Witlox

Cities, Nationalism, and Democratization
Scott A. Bollens

Life in the Megalopolis
Lucia Sa

Searching for the Just City
Peter Marcuse, James Connolly, Johannes Novy, Ingrid Olivo, Cuz Potter and Justin Steil

Forthcoming:

Urban Assemblages
How actor network theory changes urban studies
Ignacio Farias and Thomas Bender

Globalization, Violence and the Visual Culture of Cities
Christoph Lindner

Searching for the Just City

Debates in urban theory and practice

Edited by

Peter Marcuse, James Connolly, Johannes Novy, Ingrid Olivo, Cuz Potter and Justin Steil

LONDON AND NEW YORK

First published 2009
by Routledge
2 Park Square, Milton Park, Abingdon, Oxon, OX14 4RN

Simultaneously published in the USA and Canada
by Routledge
711 Third Avenue, New York, NY 10017

Routledge is an imprint of the Taylor & Francis Group, an informa business

First issued in paperback 2011

Typeset in Times NR by Graphicraft Limited, Hong Kong

British Library Cataloguing in Publication Data
A catalogue record for this book is available from the British Library

Library of Congress Cataloguing in Publication Data
Searching for the just city : debates in urban theory and practice / Peter Marcuse . . . [et al.].
p. cm.
1. City planning—Moral and ethical aspects. 2. Urban policy—Moral and ethical aspects.
3. Social justice. I. Marcuse, Peter.
HT166.S3565 2009
174′.93071216—dc22

2008045128

ISBN 10: 0-415-77613-9 (hbk)
ISBN 10: 0-415-68761-6 (pbk)
ISBN 10: 0-203-87883-3 (ebk)

ISBN 13: 978-0-415-77613-4 (hbk)
ISBN 13: 978-0-415-68761-4 (pbk)
ISBN 13: 978-0-203-87883-5 (ebk)

Contents

Figures and tables

FIGURES

TABLES

Notes on contributors

James Connolly is a doctoral student in Urban Planning at Columbia University. He has worked as a community organizer for the Austin School District and as an adviser with several community-based organizations in Texas and New York City. His dissertation research utilizes a mixed-method approach, including spatial analysis and interview methods, to analyze the role of community organizations within larger organizational fields of urban policy-making. He is looking both at the macro-historical perspective as well as at contemporary micro-patterns of community-based organizational formation and networking. He has taught urban spatial analysis for several years and his publications have focused on brownfield redevelopment processes, organizing around the Community Reinvestment Act, and on spatial patterns of advanced service firms in U.S. urban regions.

James DeFilippis is an Assistant Professor at the Edward J. Bloustein School of Planning and Public Policy at Rutgers University. His research focuses on urban political economy and political philosophy; the relationships between housing, neighborhoods, and states; community development theory and practice; and growth and proliferation of unregulated work. He has published work in academic journals in a variety of fields, both independently and in collaboration with other authors. He has also written applied monographs and reports, and his interests extend well beyond the academy and into the practice of concrete political work and policy analyses. His research has been both empirical and theoretical, and has involved quantitative and qualitative methodologies. As a policy analyst, he has worked extensively with a variety of organizations ranging from community development credit unions and Alinsky-inspired groups to broad-based housing coalitions. He is author of *Unmaking Goliath* (2004) and co-editor of *The Community Development Reader* (2007).

Mustafa Dikeç is Lecturer in Human Geography in the Geography Department at Royal Holloway, University of London, and a member of the Department's Social and Cultural Geography Group. He was trained as an urban planner at the Middle East Technical University, Ankara, Turkey, and the University of Pennsylvania, Philadelphia. He obtained his Ph.D. in urban planning from the University of California, Los Angeles in 2003. His research interests include urban theory, space and politics, social theory and space, and hospitality and the politics of alterity. He is on the editorial board of *Environment and Planning D: Society and Space* and a member of the international advisory board of *Espaces et sociétés*. His publications include numerous journal articles as well as *Badlands of the Republic: Space, Politics, and French Urban Policy* (2007).

Susan S. Fainstein is a Professor in the Urban Planning program at the Harvard University Graduate School of Design. She previously taught at Columbia and Rutgers Universities. Her teaching and research have focused on comparative urban public policy, planning theory, and urban redevelopment. Among her books are *The City Builders* and *Cities and Visitors* (co-edited with Lily M. Hoffman and Dennis R. Judd). She is the recipient of the 2004 Distinguished Educator Award of the Association of Collegiate Schools of Planning (ACSP).

Frank Fischer is Professor of Political Science and Public Administration and Faculty Fellow of The Center for Global Change and Governance at Rutgers University. He has published numerous articles on public policy and the theory and methods of social science. His published books include *Politics, Values and Public Policy* (1980), *Technocracy and the Politics of Expertise* (1990), *The Argumentative Turn in Policy Analysis and Planning* (1993), co-edited with John Forester, *Greening Environmental Policy: The Politics of a Sustainable Future* (1995), co-edited with Michael Black, *Citizens, Experts and the Environment: The Politics of Local Knowledge* (Duke University Press, 2000), and *Reframing Public Policy: Discursive Politics and Deliberative Practices* (Oxford University Press 2003). He has taught and lectured in countries throughout Europe and North America as well as South Africa and Brazil. He is on the editorial board of numerous academic journals, including *Organization and Environment*, *Administration and Society*, and the *International Journal of Public Administration*. He is winner of the 1999 Harold Lasswell Award of the Policy Studies Organization.

Ravit Goldhaber is a Postdoctorate Fellow at the Geography Department at Stellenbosch University, South Africa. Her research deals with urban segregated space, relations between social groups in the city and gated communities.

David Harvey is Distinguished Professor of Anthropology, Earth and Environmental Sciences, and History at the Graduate Center of the City University of New York. His published works include *Social Justice and the City* (1973), *The Limits to Capital* (1982), *The Urbanization of Capital* (1985), *The Condition of Postmodernity* (1989), *Justice, Nature and the Geography of Difference* (1996), *Spaces of Hope* (2000), *The New Imperialism* (2003), *Paris, Capital of*

Modernity (2003), and *A Brief History of Neoliberalism* (2005). He is recipient of the 1989 Anders Retzius Gold Medal of the Swedish Society of Anthropology and Geography, the 1995 Patron's Medal of the Royal Geographical Society of London, and the 1995 Vautrin Lud International Prize for Geography, among other awards.

Karina Leitão graduated as an architect from the Federal University of the State of Para, Brazil. She holds a Master's degree from the Latin American Integration Program at the University of São Paulo. She is currently a doctoral student in urban planning at The School of Architecture and Urbanism at the University of São Paulo-FAUUSP, and she is working as a researcher for the Housing and Human Settlements Center-LABHAB/FAUUSP.

Bruno G. Lobo is a doctoral candidate and teaching assistant at the Urban Planning program at the Graduate School of Architecture, Planning, and Preservation at Columbia University in the City of New York. He is a registered architect and urban planner and has worked as a project leader for a Lisbon-based architecture and planning firm. His research focuses on land-use regulations and property rights in a comparative perspective and its impacts on property development. Prior to his doctoral studies he earned a Master's in Architecture from the Technical University of Lisbon and was a research associate at Milan Polytechnic, Kyushu University, and the University of Tokyo.

Peter Marcuse is a planner and lawyer and is Professor Emeritus of Urban Planning at Columbia University in New York City. He has a Ph.D. in planning from the University of California at Berkeley, was Professor of Urban Planning at UCLA, and President of the Los Angeles Planning Commission and member of Community Board 9 in New York City. His fields of research include city planning, housing, the use of public space, the right to the city, social justice in the city, globalization, and urban history, with some focus on New York City. He has taught in West and East Germany, Australia, the Union of South Africa, Canada, Austria, and Brazil, and written extensively in both professional journals and the popular press. His books include *Globalizing Cities: A New Spatial Order?* (1999), with Ronald van Kempen, and *Of States and Cities: The Partitioning of Urban Space* (2002).

Erminia Maricato is a Professor in the School of Architecture at the University of São Paulo. She is the former Secretary of Housing and Urban Development of the City of São Paulo (1989/1992) and Vice Minister of Cities of Brazil (2003/ 2005). She has published widely in academic and professional journals on issues of Brazilian urbanism. She has authored several books including *Metropole Na Periferia Do Capitalismo: Ilegalidade, Desigualdade E Violencia* (1996) and *Brasil, Cidades: Alternativas Para a Crise Urbana* (2001).

Margit Mayer is Professor of Politics at the Free University of Berlin. She has also taught at the New School for Social Research and the University of California, Santa Cruz. Her current research focuses on the intersection of employment and community development policies in the United States and Germany. She has published widely in scholarly and professional journals. Her publications include *Nonprofit-Organisationen und die Transformation lokaler Beschäftigungspolitik* (2004), with V. Eick, B. Grell, and J. Sambale, and *Urban Movements in a Globalising World* (2000), with P. Hamel and H. Lustiger-Thaler (eds).

Johannes Novy is currently finishing his Ph.D. in Urban Planning at Columbia University's Graduate School for Architecture, Planning, and Preservation. He is also a researcher at the Center of Metropolitan Studies (CMS) in Berlin. Novy's research interests include planning history and theory, and urban tourism, as well as urban development in North America and Europe.

Roy Nuriel is completing his M.A. at the Geography Department at Ben-Gurion University. His research deals with the planning of multicultural Beer Sheva.

Ingrid Olivo is a doctoral student in the Urban Planning department at Columbia University. She is currently doing comparative research on the role of cultural heritage in post-disaster development planning. Born in San Salvador, El Salvador, Ingrid has a B.Sc. in Architecture (UCA El Salvador), M.Sc. in Urban Development Planning (UCL UK), and has worked as a journalist and an urban planner in two NGOs, the San Salvador Metropolitan Bureau of Planning, and two private firms. She has taught at Universidad Centroamericana Jose Simeon Canas (ES) and coordinated the set-up of a Master's Degree Program in Urban Planning at the Universidad Politecnica de Catalunya (Spain).

Cuz Potter is a doctoral student in the Urban Planning Department at Columbia University and holds Master's degrees from Columbia in international affairs and urban planning. His dissertation explores the impact of containerization's attenuation of the geographical linkage between ports and other economic activity on interurban competition. He is also employing psychoanalytic theory to investigate the role utopian thinking can play in urban planning. He has worked as an editor and translator for the South Korean Ministries of Environment and Labor in Seoul, and has consulted for a variety of entities in New York City, including the Manhattan Borough President's Office, Herrick, Feinstein, LLP, and CIVITAS. He has co-authored "The State of the Art: Regional economies, open networks and the spatial fragmentation of production" (*Socio-Economic Review*, 2007) with Josh Whitford, "Inside Informality: Poverty, Jobs, Housing and Services in

Nairobi's Slums" with Sumila Gulyani and Debu Talukdar for the World Bank, a chapter with Peter Marcuse entitled "The Heights: An Ivory Tower and Its Community" (in *University as Developer*, 2004), and "A Tale of Three Northern Manhattan Communities" with Richard Bass for the *Fordham Urban Law Journal* (2004).

Justin Steil is pursuing a J.D./Ph.D. in Law and Urban Planning at Columbia University. Most recently, he worked as community development program manager for an environmental justice organization focusing on brownfields and, before that, as advocacy director for a financial justice non-profit working against predatory lending, both in New York City. Previously, he worked with a domestic violence crisis center in Mexico, training police in the support of survivors of sexual assault, and taught at The City School in Boston, bringing prisoners and young people together to think critically about violence and justice. Broadly interested in how power is exercised through control over space, his dissertation research focuses on the intersection of law and urban planning in local efforts to regulate immigration through housing and land use controls.

J. Phillip Thompson is Professor of Urban Planning and Political Science at the Massachusetts Institute of Technology. He has worked as Deputy General Manager of the New York Housing Authority and as Director of the Mayor's Office of Housing Coordination. He is a frequent advisor to trade unions in their efforts to work with immigrant and community groups across the United States. He has published numerous journal and media articles on race, immigration, and community development in U.S. cities. He is author of *Double Trouble: Black Mayors, Black Communities and the Call for Deep Democracy* (2005).

Laura Wolf-Powers is Assistant Professor of Community and Regional Planning at the University of Pennsylvania. She has served as Chairperson of the Graduate Center for Planning and the Environment in the School of Architecture at the Pratt Institute and as a Research Fellow at the Pratt Center for Community Development. Her research and publications focus on urban labor markets, job-centered economic development, and the politics of city planning. She works actively with New York City community-based organizations on land use, workforce development, and housing initiatives.

Oren Yiftachel is a Professor of Geography and Planning at Ben-Gurion University of the Negev. He works on the political geography of planning, land, and ethnic relations. He is the founding editor of the journal *Hagar: Studies in Culture, Politics and Place* and serves on the editorial boards of: *Planning Theory* (essay editor), *Society and Space*, *IJMES*, *MERIP*, *Urban Studies*, *Journal of Planning Literature* and *Social and Cultural Geography*. He is currently working on three main research projects: the spatial transformation of Israel/Palestine, the geography of ethnocratic power structures and the various shades of the "grey city" using a comparative international perspective. He has published widely in leading academic journals. His books include *Planning and Social Control: Policy and Resistance in a Divided Society* (1995), *The Power of Planning* (2002), with D. Hedgcock, J. Little, and I. Alexander (eds), and *Ethnocracy: Land and Identity Politics in Israel/Palestine* (2006).

Preface

Peter Marcuse

I am not the senior editor of this volume, except by age. It is an actively edited volume by five very able young scholars with significant original contributions by each, plus major original essays by leading writers in the field. It is not an ordinary collection of disparate papers loosely organized around a topic, not a mildly polished conference proceeding, but rather carefully written contributions to a single complex and important theme: the meanings and use of the idea of a "Just City," both in theory and in practice. Although it did indeed have its origins in an invitational conference at Columbia University, this volume has been worked on for over two years, both as to substance and form, to make it a cohesive whole, addressing a single subject important in urban development and planning, from different viewpoints centered about reactions to a single innovative approach. The history of the book is worth setting forth:

In 2006 Susan Fainstein had been chair of the Ph.D. program in Urban Planning at Columbia University's Graduate School of Architecture, Planning, and Historic Preservation for several years, but was leaving for a different position at Harvard. She had made a very positive contribution to the program, and had worked closely and with unusual skill and understanding with our students, including, at that time, the five true editors of this volume. Following the inspiration of two of the editors of this volume, Johannes Novy and James Connolly, our program sponsored a conference in her honor around the theme of "The Just City," a concept that she had been instrumental in developing over the preceding years. The conference was by all accounts a success, and produced some very notable contributions. We decided the strongest and most fully developed among them deserved wider circulation, and in more than the usual "conference proceedings." The editors of this volume got together to undertake the task. The result is before you.

I have both edited and contributed to many edited volumes over the years, and I may say honestly I have rarely seen as engaged and active a set of editors as were involved here. Each piece, whether their own, co-written with others, or by others, was carefully reviewed, each one by all of the editors and then overseen to its final form by a one- or two-person team. Their standards were high; I must say, as I watched their comments go by me on e-mail, I sometimes held my breath at their audacity, at suggesting to much more senior scholars that this or that logic did not quite hold up, this or that needed evidence, this or that seemed internally contradictory. But their points were almost always well argued, and positively responded to; they held to a high standard of quality, and their expectations were met.

While the development of this book from the idea stage onward has been the result of a truly collaborative effort, it is nonetheless worth noting what jobs were performed by each of the five editors individually, to show specifically with what he or she should be credited (or for what conceivably blamed!).

James Connolly chiefly coordinated the editing of the publication, co-authored the Introduction ("Finding justice in the city") and the chapter "Can the Just City be built from below? Brownfields, planning, and power in the South Bronx," did the final edit on the chapters by Novy and Mayer, and by Marcuse, and on the Postscript, and participated intensively in the editing and review of all chapters.

Johannes Novy assisted in coordinating the editing of the publication, co-authored the chapter on "As 'Just' as it gets? The European City in the 'Just City,'" co-authored the Conclusion ("Just City discourse on the horizon: summing up, moving forward"), did the final edit on the chapters by Steil and Connolly, by Fischer, and by Dikeç, and participated intensively in the review of all chapters.

Cuz Potter worked with David Harvey in compiling and editing the chapter "The right to the Just City," co-authored the Conclusion ("Just City on the Horizon: summing up, moving forward"), and did the final edit on the chapters by Marcuse, by Yiftachel *et al.*, by Maricato, by Dikeç, and by Fischer, and participated intensively in the review of all chapters.

Justin Steil co-authored the chapter "Can the Just City be built from below? Brownfields, planning, and power in the South Bronx," co-authored the Introduction ("Finding justice in the city"), did the final edit on the chapters by Fainstein, Wolf-Powers, Thompson, and DeFilippis, and participated intensively in the review of all chapters.

Ingrid Olivo did the final edit on the chapter by Dikeç, and participated intensively in the review of all articles.

Whether their committed efforts have paid off the reader can judge for her or himself.

Acknowledgments

This book grew out of a conference entitled Searching for the Just City, which was held at Columbia University on April 11, 2006. The conference, sponsored by The Graduate School of Architecture, Planning and Preservation (GSAPP) at Columbia University, provided an excellent venue within which to begin the discussion of the issues covered in this volume. The editors would like to thank all of the presenters at the conference (in addition to authors in this volume, these include: Eddie Bautista, Robert Beauregard, Eugenie Birch, Diane E. Davis, Dolores Hayden, John Logan, Setha Low, John Mollenkopf, and Elliott Sclar), the additional sponsors of the conference (The Center for Urban Research and Policy, The Institute for Social and Economic Research and Policy, and the Barnard College Department of Urban Studies), the other organizers of the conference (namely, Erika Svendsen, Gabriella Carolini, Susan Gladstone, Yumie Song, Ben Prosky, and the audio visual department at GSAPP), and the conference attendants who helped to make it such a success and who provided inspiration for us to continue with the effort that it began. We would also like to thank Dean Mark Wigley of GSAPP for early and continual support for this project.

The initial proposal for this book was formulated shortly after the Searching for the Just City conference and became a central topic of concern for a semester-long doctoral colloquium on notions of justice in urban planning led by Professor Peter Marcuse. All of the members of that colloquium helped shape the issues that the book covers and the questions that it asks. They include (in addition to the editors of this volume): Shagun Mehotra, Emmanuel Pratt, Greta Goldberg, Constantine Kontokosta, Bruno Graca Lobo, and Joyce Rosenthal. We are grateful for their patient help in shaping our initial proposal. As well, Professors Susan Fainstein and Robert Beauregard provided crucial feedback and advice during the colloquium and the development of the book proposal.

We would especially like to thank Professor Marcuse for leading the colloquium and for encouraging us to take on this project and supporting us so consistently from the very beginning, often seeing the possibilities for the volume that at times escaped us. We can only imagine the incredible restraint Professor Marcuse must have exercised in turning the leadership of the editing over to our inexperienced hands and letting us learn from our mistakes as we struggled through the process. Professor Marcuse's role in the editing exemplifies his dedication and his generosity as an educator, perpetually excited about ideas of urban justice and always seeking to create opportunities for his students.

Special appreciation is also due to Professor Susan Fainstein who inspired us to organize the conference that led to this volume and who entrusted us with her chapter on the Just City when the book's future may have seemed less than certain. Throughout the editing process, Professor Fainstein has offered invaluable comments and supported us in bringing the volume together.

We would also like to thank the authors of all the chapters, without whom the book would not exist. We are indebted to them both for their original contributions and their patience as the book came together. Thanks are due as well to Andrew Mould and Michael P. Jones at Routledge for seeing the possibility in the proposal and for always offering positive words of support throughout the process.

Finally, all of the editors wish to extend their personal gratitude to those who supported them through the process of creating this book. Specifically, Johannes Novy sends his thanks to his parents, Bea Novy and Günther Uhlig, as well as Elena Blobel for their support and much appreciated perspective on life. Cuz Potter would like to thank Yoonkyung Chang, whose unflagging generosity allowed him to put his energies into this project.

James Connolly, Johannes Novy,
Ingrid Olivo, Cuz Potter, and Justin Steil

Introduction

Finding justice in the city

James Connolly and Justin Steil

The search for a Just City is, in part, an effort to realize the transformative potential of urban theory. It is a search that begins by examining the everyday reality of city life and then seeks a means to reshape that reality and re-imagine that life. It begins with the injustices that have come with rapid urbanization—the violence, insecurity, exploitation, and poverty that characterize urban life for many, as well as the physical expressions of unequal access to social, cultural, political, and economic capital that arise from intertwined divisions between race, class, and gender categories. Awareness of these everyday injustices "thickens our deliberations and provides us with a metric for evaluating our achievements" (Beauregard 2006).[1] Whether displaced merchants are challenging the dominant economic development régime, local communities are seeking to direct the remediation of contaminated urban manufacturing sites, or domestic workers are struggling for fair labor standards, actions against specific injustices provide partial, yet continual, challenges to the inequalities in everyday urban life. Awareness of exploitation, and attempts to challenge it, bring us closer to realizing the too often unfulfilled promise that cities have long represented—the promise of liberation and opportunity.[2] But to search for a Just City is to seek something more than individualized responses to specific injustices. It requires the creation of coherent frames for action and deliberation that bring the multiple and disparate efforts of those fighting against unjust urban conditions into relief and relate their struggles to each other as part of a global orchestration improvised around the single tenor of justice.

As urban residents with different backgrounds and conflicting ideologies seek to universalize their competing notions of justice, a crucial question has been the extent to which coherent and useful ideals can be imagined from this contested concept.[3] Though spatially informed notions of social justice have become a unifying cry for a number of activist organizations and urban social movements around the globe,[4] the meaning of justice for urban life remains ill-defined. The search for this meaning is, as well, the search for a Just City. This volume brings together contemporary urban theorists and practitioners in order to sharpen the definition of justice in the context of twenty-first century urbanization by engaging with the philosophic and economic conflicts that emerge from the contemporary politics of city building. Such engagement values the pluralism of ideas and the creativity that comes with sorting through these conflicts. Despite differences, the contributions to this volume all share a critique of power and exploitation, a critical reflexivity and an unwavering belief that there are more just alternatives. Bruno Latour (2004: 246) suggests that "the critic is not the one who debunks, but the one who assembles. The critic is not the one who lifts the rugs from under the feet of the naïve believers, but the one who offers the participants arenas in which to gather." This volume assembles critics of existing urban injustice and offers them an arena in which to gather, taking the concept of the Just City as developed by Susan Fainstein as a starting point for assembling a theoretical frame that can effectively direct deliberation and action in the process of reshaping existing urban realities. The conclusion and postscript of this volume offer an assessment of where we are and where we need to go next in this effort. To begin, though, some clarification on where we have been with regard to the literature on justice and the city is required.

APPROACHES TO DEFINING JUSTICE

The city has always been a fundamental heuristic within attempts to define justice. In Plato's *The Republic*, Socrates uses a lengthy description of what a Just City would look like as scaffolding for his argument in favor of justice as an ethical guide for individuals. He argues that the qualities of justice are more easily discerned within the actions of the State (specifically the city-state in the context of Plato's time) than they are from within those of the individual. Thus, in order to respond to claims by Thrasymachus that justice is simply an expression of what is best for those in power and therefore, that the unjust life should be preferred to the just, Socrates creates a Just City in words. Socrates' Just City, and the definition of justice it entails, necessarily engages political questions concerned specifically with the distribution of power in order to respond to Thrasymachus' claims (see Neu 1971; Lycos 1987). Justice, in this context, is defined as an internal quality of cities based upon the aggregated actions of individual residents—in some ways, an expression of the situated nature of justice echoed by several contemporary urban theorists discussed below. Ultimately, though, the parameters of a Just City are not so much specified within *The Republic* as they are set in contrast to those of existing cities of the time. Socrates ends by prescribing a much contested "organic model" for the State that entails a strong and growing role for the public sector (which Socrates envisioned as populated by political philosophers), in order to ensure a Just City and, thus, a just life for its inhabitants (see Dobbs 1994).

Liberal political philosophy

This initial concern for the distribution of power and the role of the State in creating a Just City is reflected (though not always specifically addressed) in Western debates about justice that have developed over the last half-century, mostly within the fields of political philosophy and political economy. Contemporary definitions of justice are generally seen to fall under the domain of liberal political philosophers, from John Locke, Jean-Jacques Rousseau, and John Stuart Mill to John Rawls and Martha Nussbaum. Rawls' *A Theory of Justice* (1971) and his idea of the "original position" in which individuals choose principles of justice from behind a "veil of ignorance" emphasizes the value of liberty and equality. Rawls argues that everyone has an equal right to basic liberties and that social and economic inequalities, where necessary, should be distributed to benefit the least advantaged. His advancement of a normative social contract theory critiques Mill's utilitarianism[5] and, when it was published, reinvigorated liberal political philosophy in the Anglo-American context. It conceives of justice in abstract and universal terms, separated from existing political contexts but, at the same time, accepts much of the capitalist economic structure. Amartya Sen (1999) and Martha Nussbaum (2000) build on the social contract theory of liberal political philosophy to emphasize a more detailed capabilities approach to understanding justice. These capabilities, which include the ability to maintain political and material control over one's environment, represent nontradeable, inalienable opportunities to which each person should be entitled. Like the Western liberalism from which it emerges, the capabilities approach largely treats individuals as abstract, universal, atomistic actors disembedded from their social relations and historical and spatial specificities. This account of essential human functionings and rights thus fails to fully come to terms with the importance of the situatedness of both author and subject and the implications that difference has for people's everyday lives, needs, and wants.

Neither Rawls, Sen nor Nussbaum elaborate how their normative conceptions of justice, based on equality and fairness, can be realized or what forms they might take, a problem that has characterized the philosophy of justice since Socrates' attempt to define a Just City in *The Republic*. All three recent formulations leave readers questioning what justice might mean as a concrete structure within everyday life. Such detail on what a Just City would look like in the context of the modern State is not a question these philosophically oriented political theorists seek to answer, though they offer important elaborations on the complexity inherent in the concept.

Communicative rationality

Another influential approach to justice that emerged within liberal political philosophy since the end of the Second World War and that has been a theoretical base for many seeking to create a more just urban form is the idea of communicative rationality, articulated most prominently by Jürgen Habermas (1985). Building on the tradition of pragmatism, and his own work on the public sphere (1962/1991), Habermas emphasizes the importance of discourse ethics and the "ideal speech situation" for creating discursive theories of democracy. This emphasis on discourse and social relations allows for more historically and spatially situated understandings of justice while avoiding total relativism. With the rise of postmodern challenges to grand narratives, this focus on the processes that could lead to justice, instead of a definition of its ends, has become increasingly popular and led, among other things, to a school of communicative urban planning practice in which discursively democratic means constitute the path to a just end (Healey 1997; Forester 1999). This equation for urban justice, largely formulated as a postmodern defense of the "cultural politics" of difference (see Soja 1997: 184), has been criticized for failing to recognize the impossibility of creating truly ideal speech situations in the context of drastic political and economic inequality and the reality that unjust ends can result from relatively just processes. The implicit meaning of this approach for urban theory, then, is that the ends are left to be the *de facto* result of practice while only the means are theorized, a position further critiqued by authors in this volume.

Political economy

Recognition of the uneven power positions that complicate the ideals of discursive planning along with a historically and economically situated critique of political philosophy are at the heart of the political economy approach to justice. Karl Marx dismissed liberal conceptualizations of justice as *bourgeois* prejudices hiding *bourgeois* interests. Many of those inspired by Marx continue to emphasize the need to focus on the very concrete inequalities continually reproduced by capitalist modes of production and accumulation. David Harvey's seminal work *Social Justice and the City* (1973) begins by trying to analyze urban problems from a Rawlsian liberal perspective, but fails to find satisfactory answers in this realm. Turning to a Marxist analysis, Harvey identifies unequal spatial development as fundamental to the functioning of capitalism. Instead of confronting the symptoms visible in urban decline, Harvey argues that justice demands the transformation of the processes that gave rise to urban inequality in the first place—the asymmetries of economic and political power embedded in the practices of capital accumulation. In the end, Harvey calls for the exploration of alternative modes of production, consumption, and distribution that would reorganize the class structure of society. Such an approach, common amongst Marxian urban theory of the time that *Social Justice and the City* was published (see also Castells 1977, 1978, 1983; Katznelson 1982), has been critiqued for its supposition that class is a single unifying category with the ability to universalize the particularities of other identities such as race, ethnicity, and gender (see Tajbakhsh 2001). As well, the political economic approach, including that of some early Marxian urbanists, has evolved in recent decades both along with and in response to a poststructuralist understanding of social systems.

This evolution, which in part is characterized by a "cultural turn" in the political economic perspective, has been shaped by a conscious effort to examine the inequalities of uneven development across a wider spectrum than the Marxian focus on class. Influenced by postmodernist epistemology and a growing body of theories on the social production of space, this perspective in urban political economy has looked toward race, ethnicity, gender, sexuality, and other social groupings as essential markers in the uneven distribution of power and resources (see Soja 1999). In this vein, authors such as Iris Marion Young (1990, 2000) and Nancy Fraser (1999) articulate the "risks involved both in attending to and ignoring difference" (Young 1990: 86). They use empirical case studies to point out the limitations of a strictly redistributive model of social justice and highlight the crucial role of recognition. Conceptualizing social categories as produced through both material and discursive power relations, scholars working in this vein argue for the importance of recognizing claims asserted from the specificity of social group positions in order to challenge structural inequalities. Justice, from this perspective, requires not simply formal inclusion or equality but "attending to the social relations that differently position people and condition their experiences, opportunities and knowledge of the society" (Young 2000: 83). While the focus on recognition that has come with the "cultural turn" has done much to advance the interests of under-represented groups, Michael Storper (2001) highlights the attendant possibility for relativism that creates an inability to navigate conflict when group differences collide. The inclusion of recognition and difference continues to guide research, but remains a source of persistent tension amongst political economic analysts wary of moving too far away from the question of power in capitalist society.

CONTEMPORARY FORMULATIONS OF JUSTICE AND URBAN SPACE

Much has changed in global politics since Harvey wrote *Social Justice and the City* in 1973 and since the advent of poststructuralist calls for recognition in the 1990s, but the rise of neoliberalism to hegemonic status has only heightened the unevenness of spatial development. Government initiatives to deregulate financial and other industries, privatize public goods, restrain unions and limit workers' rights have been combined with efforts to strengthen private property rights and extend free trade and market incentives to new economic sectors and new global regions (Harvey 2005), a trend heightened by current responses to the global economic recession. Within contemporary urban politics, the dominance of neoliberalism and discourses of the competitive city have effectively redirected attention away from traditional issues of social justice and toward a new liberal formulation of social problems as questions of "social cohesion, social exclusion and social capital" (Harloe 2001: 890), but this turn has also been met with resistance at the local and global level.

Movements against neoliberalism, like all liberation movements, are both struggles in space and also struggles for space (Merrifield and Swyngedouw

1997). Understanding justice requires not only engaging with the dialectical relationship between social and economic conditions, but also with the spatial implications of that relationship. How does attention to the production and experience of urban space illuminate the philosophical articulations and concrete struggles for social justice? As Smith (1992), Purcell (2003), Brenner (2004), and others have pointed out, the rescaling of governance has brought renewed attention to the multiple levels at which politics and economy are negotiated. This rescaling has emphasized the significance of the municipal level in struggles over neoliberalism and social justice. Arguably, the city is the scale large enough for a government to have meaningful power, but still small enough for a democracy in which people can actually affect politics (see Dahl 1967 for an early discussion of this topic; see also Fainstein this volume; DeFilippis this volume). It is in cities that the place of community is organized relative to the space of capital investment and that the effect upon urban residents—be they in the rapidly expanding cities of developing countries or in the postindustrial regions of the advanced economy—is ultimately decided. Actions taken at all scales of governance are certainly pertinent, but the city is the scale where questions of justice are felt concretely as part of everyday life.

Considering scale, any theory of urban justice must wrestle with the extent to which a just arrangement for those within one city's borders could coexist with, or depend upon, unequal or exploitative relations with inhabitants of other cities and non-urban areas. Much of the writing in this volume focuses primarily on North America and Western Europe. The world's fastest growing cities, and the majority of the urban population, live outside these regions in very different urban contexts. Several chapters in this volume point to the imperative of considering the Just City arguments in light of non-Western urban contexts and of learning from the innovations of "ordinary cities" and the syncretization and creolization that increasingly characterize everyday urban life around the world (see Maricato this volume; Thompson this volume; Mayer and Novy this volume; and Yiftachel *et al.* this volume; as well as King 1996; Amin and Graham 1997; Robinson 2006, *inter alia*). Attention to non-Western cities raises the history of imperialism that has long structured global urban relations (see Maricato this volume; Thompson this volume). Neoliberalism heightens the continuing significance of empire, race, and migration in shaping urban development while also creating new opportunities for finding solidarity across national borders.

In this context of neoliberal restructuring, renewed debates about justice, utopian thought, and the "right to the city" have highlighted the need for contributions that bring political philosophic and political economic understandings of justice together. Susan Fainstein has developed this line of thought in a series of articles that seek to specify a model for urban planners that "reacts to the social and spatial inequality engendered by capitalism" (Fainstein 2005: 2). Fainstein's modified form of political economic analysis, which she labels "Just City" planning,[6] takes the normative stance of political economists that favors social equity and seeks to overlay a detailed outline of the values rooted in philosophical notions of justice that guide the creation of the "good city."[7] In so doing, her theory of the Just City attempts to provide an alternative to both process-oriented paradigms of urban planning based upon Habermasian communicative rationality (i.e., the Communicative Model) and product-oriented paradigms based upon a physically determinist view of urban social life (i.e., the New Urbanism) (Fainstein 2000). As well, within her development of the Just City model, Fainstein has been simultaneously critical of postmodern calls for diversity as an unquestioned orthodoxy in city planning as well as of the Marxian use of class as a "cross cutting category over other forms of social difference."[8] For Fainstein, the values that guide the creation of public space, housing, economic development, and social programs that should exist within a Just City can and must be made explicit in order to mobilize broad-based and inclusive movements for change rooted in social rationality and a definition of the collective good.

The Just City articulation that Fainstein presents perhaps shares the most with the recent writings of Heather Campbell (2006). Campbell explores the concept of justice in the practice of urban planning, especially with regard to the role of situated ethical judgment in connecting abstract principles to concrete cases, especially in contentious circumstances. Campbell argues for the importance of a relational understanding of planning that focuses on the interdependence of individuals and communities (Campbell 2006: 101). In that relational understanding, the crucial practice that links reasoning with justice and negotiates between the universal and the particular is the exercise of judgment informed by a contextual understanding of the values at stake and the divergent perspectives involved (Campbell 2006: 102–3). To the contributions of Campbell's more philosophically rooted work, Fainstein's writing adds significant attention to political economy and to the context in which a concept of the Just City must struggle to take root.

The writings of Fainstein and Campbell continue in a long line of urban planning literature that points urban theory and practice in the direction of justice, from Friedrich Engels' (1872) "The Housing Question" and Ebenezer Howard's (1898) "To-morrow: A Peaceful Path to Real Reform" to the articulations of advocacy (Davidoff 1965) and equity (Krumholz 1982) planning theorists of the 1960s. All of these works in many ways set out to answer the questions that Fainstein has repeatedly posed for urban theory: (1) under what conditions can conscious action produce a better city for all citizens, and (2) how do we evaluate what outcomes would truly be better? These questions are also in line with the more recent reframing of social movement struggles around justice, as opposed to equality, which allows goals to be framed simultaneously in material (economic redistribution) and non-material (capabilities, opportunities, liberties) terms.

In this vein, several contemporary theorists have crafted descriptions of the "good city" in order to sketch an understanding of the relationship between justice, responsibility, and the urban (inter alia Harvey 1992, 1996; Merrifield and Swyngedouw 1997; Friedmann 2000; Amin 2006). Ash Amin highlights the centrality of "an urban ethic imagined as an ever-widening habit of solidarity" based around the concepts of "repair," "relatedness," "rights," and "re-enchantment" that are part of everyday experience (Amin 2006: 1012). Amin imagines a well-functioning, inclusive and participatory city that "celebrates the aspects of urban life from which spring the hopes and rewards of association and sociality" (Amin 2006: 1019). In another exploration of the concept, John Friedmann (2000) focuses on the very concrete issues of housing, health care, wages, and social welfare as the four pillars of the good city. Friedmann emphasizes the central role of civil society organizations in struggling to reinforce these pillars in the context of democratic institutions. The Just City concept, as articulated by Fainstein, shares much with these and other articulations of the "good city," particularly an emphasis on broadening a sense of solidarity among urban dwellers and imagining the contexts in which urban conditions can be changed for the better.

Another related and increasingly influential approach to understanding justice and the city has found inspiration in the work of Henri Lefebvre (*inter alia* Dikeç 2001; Mitchell 2003; Purcell 2003; Smith 2003). Lefebvre's provocative writings on the production of space (1992), the "urban revolution" (2003), and the "right to the city" (1996) have increasingly been a motivating "cry and a demand" (Lefebvre 1996: 158) both for scholars and for activists and organizers. Purcell (2003) has focused on the trends redefining the liberal-democratic conception of citizenship and proposes a "right to the global city" which encompasses a right both to appropriate urban space and to produce it through participation in decision-making at all the scales that affect the inhabitant. The emphasis on the concept of a right to the city, a right both to use it and to participate in its social and political production, has animated a dynamic coalition of community organizations and other civil society groups across the U.S. calling for economic and environmental justice. Members of this Right to the City Alliance have been active nationwide fighting gentrification and calling for a right to land and housing free from the pressures of real estate speculation and that can serve as cultural and political spaces to build sustainable communities. The Postscript to this volume takes up some of the implications of the right to the city concept for further theoretical and practical work.

The right to the city, Just City, and good city formulations share a desire to rearticulate the political and moral connections between inhabitance, social provision and social justice (Ong 2006). Dikeç (2001) articulates the shared goal within of this scholarly work—that is, the development of "a conceptual apparatus that could be given normative content to guide the actual production of urban space" (Dikeç 2001: 1803)[9]—and the differing set of theoretical criteria from which these perspectives draw. For Dikeç, the challenge is to articulate a right to difference and a right to the city within the spatial dialectics of injustice. While Fainstein's work does not preclude such a focus, she looks toward an explicit engagement with the philosophical and political foundations that can justify criteria and visions of a good city. A strong philosophical grounding and carefully argued justifications, Fainstein suggests, are crucial for any effort to widen feelings of solidarity, mobilize civil society and effectively motivate a broad base of actors to overcome existing social divisions and press for more progressive urban policies. Dikeç's theoretical mechanism, grounded in the right to the city concept, highlights, on the other hand, the role of emancipatory politics in mobilizing the marginalized rather than initially seeking to create a broader-based mobilization in the name of spatial justice. While the points of intersection between these perspectives are strong, the points of departure are important for urban politics. For this reason, Dikeç's work is substantially reprinted in this volume as a means of highlighting some of the similarities and differences between the Just City, good city, and right to the city perspectives.

AN ARENA FOR THE JUST CITY: CRITIQUES AND DEBATES THAT ADVANCE THE DISCUSSION

To create the arena described by Bruno Latour in which critical scholars can gather, this book brings together authors from a variety of disciplines to explore the potential—and tensions— embodied in the concept of the Just City. In this exploration, the authors' diverse perspectives help to illuminate the relation between the roots of urban injustices and the vision required to respond to those injustices. Categorical divides between schools of urban theory must be addressed and reconciled to some degree to unite actors in processes of both action and deliberation. The authors in this book seek such an engagement and, in providing an active example of critical reflection and interdisciplinary dialogue, implicitly invite readers to do the same.

The book is organized in three parts that correspond roughly to a scale of analysis and a set of distinct, but interrelated, questions. The first part addresses fundamental philosophical and political tensions in the discussion. The chapters in the second part examine closely a select set of contested aspects of those debates and expand the boundaries of the discussion in light of applied planning practices. Those in the final part look at concrete case studies and their implications. The three groups of questions with which this volume engages are:

Part I: *Why Justice? Theoretical Foundations of the Just City Debate:* Can justice be defined positively or must it be expressed only as the absence of injustice? Either way, can a Just City be universally articulated or does it depend fundamentally on positionality and local specificities? If it can be affirmatively defined, what would constitute "justice" in an urban setting?

Section 2: *What are the Limits of the Just City? Expanding the Debate:* Is the city level the appropriate scale for analysis and action by which to challenge contemporary processes of urban and regional development? To what extent can justice be defined within the context of existing societal relations without inherently reflecting the interests of the stronger?

Section 3: *How Do We Realize Just Cities? From Debate to Action:* Can the concept of the Just City actually stimulate visions of a better society? If so, how are such visions useful in practice? If the concept of justice is not a useful measure by which to guide and evaluate urban development, what should take its place? What social, political, economic, and other structures are needed to produce a Just City?

In Part I, Susan Fainstein works from a political economic lens to contextualize philosophical conceptions of justice within the major theoretical tensions to which the Just City project must respond. David Harvey critiques such a formulation of the Just City from a Marxist perspective, arguing that any attempt to realize justice within the context of capitalist relations will fail to address the root causes of injustice. Frank Fischer critiques the project from a deliberative democratic perspective rooted in the discursive paradigm of planning and public policy, but also seeks to find common ground between the political economic and discursive approaches. Finally, Mustafa Dikeç draws on Lefebvre to focus attention on the right to difference and the central role of those most marginalized in finding a route to a Just City.

Part II is about expanding the boundaries of the Just City debate. These authors question whether the Just City as articulated by Fainstein goes "far enough" and examine the complexity associated with using justice as a primary evaluation tool in cities. Peter Marcuse calls for these boundaries to be expanded in a structural sense by looking beyond calls for simple distributional equity to a wider delineation of "commons planning" as opposed to "justice planning." Johannes Novy and Margit Mayer call for the boundaries to be expanded in a geographic sense. They seek to include cities outside of the U.S. and Europe in the discussion, questioning the choice of Amsterdam as a model for a Just City and critiquing Fainstein's acceptance of growth criteria based on their assessment of developments in European urbanism. Analyzing urban policy in Israel, Oren Yiftachel, Ravit Goldhaber, and Roy Nuriel seek to expand the notions of difference and recognition that are used in developing a Just City. They show how "affirmative recognition," "marginalizing indifference," and "hostile recognition" for different ethnic groups has created a "creeping urban apartheid" and "gray space" in Israeli cities. Finally, James DeFilippis seeks to expand the issues being examined by, somewhat paradoxically, shrinking the language of globalization and re-infusing the local into economic development discussions. Drawing from research on largely unregulated employment sectors, he highlights the localized nature of most capital–labor relations and the need to refocus our attention from the threat of competition between global cities to the reality of economic exploitation within cities and workers' struggles.

The third section turns to the practical matters of achieving just outcomes on the ground. It asks what urban actors seeking to realize the goals of the Just City do and what challenges they face. Laura Wolf-Powers highlights the importance of public discourse as an active force in creating "counter-publics." She shows how, in the case of Brooklyn, New York in the 1960s and 1970s, alternative discourses reframed urban issues in ways that changed how neighborhood residents felt about their own efficacy and was crucial for progressive planners and activists to gain traction in broader policy discussions. Justin Steil and James Connolly examine issues of institutional structure and just urban outcomes through a case study of grassroots environmental justice organizations in the South Bronx section of New York City. These community-based organizations have been working to reconfigure the organizational relations of brownfield redevelopment in order to create a more multilateral structure that includes insurgent voices. Looking at similar issues at the national scale, Erminia Maricato analyzes Brazil's "right to the city" inspired legislation and its aim of restructuring property rights and social welfare provision, as well as the obstacles to its realization. Finally, addressing the very American disaster of New Orleans, Phil Thompson shows how any effort to create a Just City in the U.S. must recognize the extent to which politics and power are inextricably tied to the history of slavery and colonialism and continuing race-based labor exploitation.

The volume concludes with Johannes Novy and Cuz Potter looking again at the issues that the volume was chiefly concerned with and evaluating where the chapter authors' contributions have brought us. Peter Marcuse has the last word in a Postscript that highlights the changes that the civil rights movement and the uprisings in 1968 brought to our understandings of socially just goals. Marcuse points towards the momentum generated by the "Right to the City Alliance" as advancing the aims articulated by those searching for just cities.

The debates in this volume are expected to both expand the scope of urban imagination and to help reinvigorate, unify, and empower shared desires for just urban outcomes. Lefebvre characterized himself as a "utopian . . . a partisan of possibilities" (cited in Pinder 2006: 239), and David Pinder and other scholars have recently emphasized again the value of progressive utopian visions for everyday urban life. In a context where neoliberalism has gained such power essentially as "a utopia of unlimited exploitation" and "a program for destroying collective structures which may impede the pure market logic" (Bourdieu 1998: 1), progressive utopian visions can emphasize the value collective structures create and open new perspectives on what urban life can become. In presenting the Just City, Fainstein argues that "while utopian ideals provide goals toward which to aspire and inspiration by which to mobilize a constituency, they do not offer a strategy for transition within given historical circumstances" and thus she seeks to develop a vision of what is desirable and feasible within the conditions in which present-day cities find themselves embedded. Both more utopian and more pragmatic approaches are necessary in efforts to realize more just cities and dialogue between the two will undoubtedly prove fruitful. As Fredric Jameson has written, "the question of Utopia would seem to be a crucial test of what is left of our capacity to imagine change at all" (1991: xvi).

On both the pragmatic and the utopian levels, the discussion of a Just City inherently raises essential questions about the geography of our responsibilities (Massey 2004)—about which actors at what scales urban inhabitants can hold accountable for the quality of their everyday life, their access to spaces and to opportunities. The concept of a Just City can create a moral and political lever that social movements can use to argue for changes in the relation between the state and the market (Yuval-Davis 1999; Brodie 2007). Paired with articulations of the right to use and participate in the production of urban space, formulations of a Just City can empower urban residents to more effectively make claims about access to space and the provision of collective resources.

Some contributors approach justice in the city by first highlighting injustice, while others examine visions of equitable alternatives. Both approaches sharpen our analyses of the processes creating injustice and challenge us to imagine unseen possibilities. The contributors to this volume help illuminate the choices that are being made every day about the form and social processes of the city. Protesting the lack of accountability of the World Trade Organization at its meeting in Seattle in 1999, demonstrators highlighted the need for broader participation in decision making and more attention to needs of the marginalized, rallying around the cry, "This is what democracy looks like!" What does just space, a Just City look like? Recognizing that cities are key sites in the reproduction of social relations of domination, this volume is an entry into the debate about what justice means in the context of the twenty-first-century city and what alternative urban futures could look like.

NOTES

1 This quote was taken from a talk presented at the Searching for the Just City conference (April 29 2006, Columbia University, New York City), which highlighted a key point of debate at the conference.

2 Of course, at least since Simmel (1903) raised the issue within the context of modern urban life, the issue of the effect of metropolitan living upon individual freedom and opportunity has been contested especially amongst sociologists. This debate notwithstanding, current shifts toward a majority urbanized global population demonstrate that the "promise" of urban life remains a large draw for people worldwide. See also the introduction by Loretta Lees in The Emancipatory City (2004).

3 Friedrich Hayek, as a key example of the conflicting ideologies at work here, argued from a decidedly pro-capitalist position that social justice is a term of art with

limited operational meaning. He further argued that the vagueness of the term's meaning is maintained by those who utilize it in order to make it a tool for coercion in the name of utopian goals (Hayek 1978).

4 See Dikeç (this volume) and Yiftachel *et al.* (this volume); see also Soureli's (2008) presentation at the American Collegiate Schools of Planning conference; see also Steil and Connolly (this volume) for a discussion of the growth of a large-scale environmental justice movement.
5 Utilitarianism as originally outlined by John Stuart Mill (1861/1969) is (and was in Rawls' time) the dominant theoretical model for distributive justice, which generally upholds "free market principles" as a mechanism for providing the greatest good for the greatest number.
6 See Fainstein (2000) for a description of "Just City" planning relative to other dominant paradigms.
7 For more on discussions of the "good city" see Harvey (1992) and (1996); Merrifield and Swyngedouw (1997); Friedmann (2000); Amin (2006).
8 Quoted from transcripts of a talk presented at the Searching for the Just City conference, April 29, 2006 at Columbia University, New York City.
9 This goal is very near to Fainstein's goal, which she informally stated as: to "provide a rhetorical method whereby my side [i.e. her normative position] might be able to win." Quoted from transcripts of a talk presented at the Searching for the Just City conference, April 29, 2006 at Columbia University, New York City.

References

Amin, A. (2006) "The Good City," *Urban Studies*, 43: 1009–1023.
Amin, A. and Graham, S. (1997) "The Ordinary City," *Transactions of the Institute of British Geographers*, 22(4): 411–429.
Beauregard, R. (2006) "Injustice and the City," presented at the Searching for the Just City conference, April 29: Columbia University, New York City.
Bourdieu, P. (1998) "Utopia of Endless Exploitation: The Essence of Neo-liberalism," *Le Monde Diplomatique*.
Brenner, N. (2004) *New State Spaces: Urban Governance and the Rescaling of Statehood*, Oxford: Oxford University Press.
Brodie, J. (2007) "Reforming Social Justice in Neoliberal Times," *Studies in Social Justice*, 1(2): 93–107.
Campbell, H. (2006) "Just Planning: The Art of Situated Ethical Judgement," *Journal of Planning Education and Research*, 26: 92–106.
Castells, M. (1977) *The Urban Question: A Marxist Approach*, Cambridge, MA: MIT Press.
—— (1978) *City, Class, and Power*, trans. Lebas, E., New York: St. Martin's.
—— (1983) *The City and the Grassroots*, Berkeley: University of California Press.
Dahl, R. (1967) "The City in the Future of Democracy," *American Political Science Review*, 61(4): 953–970.
Davidoff, P. (1965) "Advocacy and Pluralism in Planning," *Journal of the American Institute of Planners*, 31: 277–96.
Dikeç, M. (2001) "Justice and the Spatial Imagination," *Environment and Planning A*, 33: 1785–1805.
Dobbs, D. (1994) "Choosing Justice: Socrates' Model City and the Practice of Dialectic," *The American Political Science Review*, 88(2): 263–77.
Engels, F. (1872/1988) "The Housing Question," in K. Marx and F. Engels *Collected Works*, vol. 23, London: Lawrence and Wishart.

Fainstein, S. (2000) "New Directions in Planning Theory," *Urban Affairs Review*, 35(4): 451–478.

—— (2005) "Cities and Diversity: Should We Want It? Can We Plan for It?," *Urban Affairs Review*, 41(1): 3–19.

Forester, J. (1999) "Reflections on the Future Understanding of Planning Practice," *International Planning Studies*, 4(2): 175–193.

Fraser, N. (1999) "Social justice in the age of identity politics: redistribution, recognition and participation," in Ray, L. and Sayer, R. (eds) *Culture and Economy After the Cultural Turn*, London: Sage.

Friedmann, J. (2000) "The Good City: In Defense of Utopian Thinking," *International Journal of Urban and Regional Research*, 24(2): 460–472.

Habermas, J. (1985) *The Theory of Communicative Action, Volume 1: Reason and the Rationalization of Society*, Boston: Beacon Press.

—— (1991) *The Structural Transformation of the Public Sphere: An Inquiry into a Category of Bourgeois Society*, Cambridge, MA: The MIT Press.

Harloe, M. (2001) "Social Justice and the City: The New 'Liberal Formulation,'" *International Journal of Urban and Regional Research*, 25(4): 889–97.

Harvey, D. (1973) *Social Justice and the City*, Oxford: Blackwell.

—— (1992) *The Condition of Postmodernity: An Enquiry into the Origins of Cultural Change*, Oxford: Blackwell.

—— (1996) "On Planning the Ideology of Planning," in Fainstein, S. and Campbell, S. (eds) *Readings in Planning Theory*, Oxford: Blackwell.

—— (2005) *A Brief History of Neoliberalism*, Oxford: Oxford University Press.

Hayek, F.A. (1978) *Law, Legislation and Liberty, Volume 3: The Political Order of a Free People*, Chicago: University of Chicago Press.

Healey, P. (1997) *Collaborative Planning: Shaping Places in Fragmented Societies*, Vancouver: University of British Columbia Press.

Howard, E. (1898/2003). *To-morrow: A Peaceful Path to Real Reform*, London: Routledge.

Howard, E. and Osborn, F. (1898) *Garden Cities of Tomorrow*, London: Faber & Faber.

Jameson, F. (1991) *Postmodernism, or The Cultural Logic of Late Capitalism*, Durham: Duke University Press.

Katznelson, I. (1982) *City Trenches: Urban Politics and the Patterning of Class in the United States*, Chicago: University of Chicago Press.

King, A. (1996) *Re-presenting the City. Ethnicity, Capital and Culture in the 21st Century Metropolis*, London: Macmillan.

Latour, B. (2004) "Why Has Critique Run Out of Steam? From Matters of Fact to Matters of Concern," *Critical Inquiry*, 30(2): 225–248.

Krumholz, N. (1982) "A Retrospective View of Equity Planning: Cleveland, 1969–1979," *Journal of the American Planning Association*, 48(2): 163–174.

Lees, L. (2004) The Emancipatory City: Urban (Re)Visions, in L. Lees (ed.), *The Emancipatory City: Paradoxes and Possibilities*, London: Sage Publications.

Lefebvre, H. (1992) *The Production of Space*, Oxford: Blackwell.

—— (1996) *Writings on Cities*, Oxford: Blackwell.

—— (2003) *The Urban Revolution*, trans. Robert Bononno, Minneapolis: University of Minnesota Press.

Lycos, K. (1987) *Plato on Justice and Power: reading Book I of Plato's The Republic*, Albany, NY: State University of New York Press.

Massey, D. (2004) "Geographies of Responsibility," *Geografiska Annaler B* 86B(1): 5–18.
McLeod, M. (1997) "Henri Lefebvre's Critique of Everyday Life" in S. Harris and D. Berke (eds) *Architecture of the Everyday*, New York: Princeton Architectural Press.
Merrifield, A. and Swyngedouw, E. (1997) *The Urbanization of Injustice*, New York: NYU Press.
Mill, J. (1861/1969) "Utilitarianism," in *Collected Works of John Stuart Mill*, vol. X, Toronto: University of Toronto Press.
Mitchell, D. (2003) *The Right to the City: Social Justice and the Fight for Public Space*, New York: The Guilford Press.
Neu, J. (1971) "Plato's Analogy of State and Individual: 'The Republic' and the Organic Theory of the State," *Philosophy*, 46(177): 238–254.
Nussbaum, M. (2000) *Women and Human Development: The Capabilities Approach*, Cambridge: Cambridge University Press.
Ong, A. (1999) *Flexible Citizenship: The Cultural Logics of Transnationality*, Durham: Duke University Press.
—— (2006) *Neoliberalism as Exception: Mutations in Citizenship and Sovereignty*, Durham: Duke University Press.
Pinder, D. (2006) *Visions of the City: Utopianism, Power and Politics in Twentieth Century Urbanism*, London: Routledge.
Plato (1963) *The Republic*, Cambridge, MA: Harvard University Press.
Purcell, M. (2003) "Citizenship and the Right to the Global City: Reimagining the Capitalist World Order," *International Journal of Urban and Regional Research*, 273: 56–90.
Rawls, J. (1971) *A Theory of Justice*, Cambridge, MA: Harvard University Press.
Robinson, J. (2006) *Ordinary Cities: Between Modernity and Development*, London: Routledge.
Sen, A. (1999) *Development as Freedom*, New York: Alfred A. Knopf.
Simmel, G. (1903) "The Metropolis and Mental Life," trans. in Wolf, K. (ed.) (1950) *The Sociology of Georg Simmmel*, Chicago: University of Chicago Press.
Smith, N. (1992) "Geography, Difference and the Politics of Scale," in Doherty, J. Graham, E., and Malek, M. (eds) *Postmodernism and the Social Sciences.* London: Macmillan.
—— (2003) "Foreword," in Lefebvre, H., *The Urban Revolution*, Minneapolis: University of Minnesota Press.
Soja, E, (1997) "Margin/Alia: Social Justice and the New Cultural Politics," in Merrifield, A. and Swyngedouw, E. (eds) *The Urbanization of Injustice*, New York: New York University Press.
—— (1999) "In Different Spaces: The Cultural Turn in Urban and Regional Political Economy," *European Planning Studies*, 71(1): 65–75.
Soureli, Konstantina (2008) "Towards Critical Spatial Possibilities," presentation at the Association of Collegiate Schools of Planning conference, Chicago, July 6–11.
Storper, M. (2001) "The Poverty of Radical Theory Today: From the False Promises of Marxism to the Mirage of the Cultural Turn," *International Journal of Urban and Regional Research*, 25(1): 155–179.
Tajbakhsh, K. (2001) *The Promise of the City: Space, Identity and Politics in Contemporary Social Thought*, Berkeley: University of California Press.

Young, I.M. (1990) *Justice and the Politics of Difference*, Princeton, NJ: Princeton University Press.
—— (2000) *Inclusion and Democracy*, Oxford: Oxford University Press.
Yuval-Davis, N. (1999) "The Multi-Layered Citizen: Citizenship at the Age of "Glocalization," *International Feminist Journal of Politics*, 1: 119–136.

Part I

Why justice?

Theoretical foundations of the Just City debate

1 Planning and the Just City[1]

Susan S. Fainstein

The profession of city planning was born of a vision of the good city. Its roots lie in the nineteenth-century radicalism of Ebenezer Howard and his associates, in Baron Haussmann's conception of creative destruction, and in the more conventional ideas of the urban progressives in the United States and their technocratic European counterparts. While the three approaches differed in their orientation toward democracy, in their content, and in their distributional outcomes, they all had their start in a revulsion at the chaotic and unhealthful character of the industrial city. Their common purpose was to achieve efficiency, order, and beauty through the imposition of reason (Scott 1998).

Today planning is characterized by greater modesty. Despite some exceptions, especially the advocates for the new urbanism, most planners and academic commentators believe that visionaries should not impose their views upon the public.[2] Moreover, skepticism reigns over the possibility of identifying a model of a good city. Attacks on the visionary approach have come from across the ideological spectrum. The left has attacked planning for its class bias (Gans 1968; Harvey 1978), for its anti-democratic character (Davidoff and Reiner 1962; Yiftachel 1998; Purcell 2008), and for its failure to take account of difference (Thomas 1996). The right sees planning as denying freedom (Hayek 1944) and producing inefficiency (Anderson 1964); it regards markets as the appropriate allocators of urban space (Klosterman 1985). Centrists consider comprehensive planning inherently undemocratic and unattainable (Altshuler 1965; Lindblom 1959), seeing the modernists' efforts to redesign cities as destructive of the urban fabric and indifferent to people's comfort and desires (Jacobs 1961; Hall 2002). Indeed the history of planning practice seems to validate the critics: postwar American urban renewal and highway building programs resulted in displacement and the break-up of communities, while European social housing development frequently produced unattractive, socially homogeneous projects. Now, the emphasis on economic competitiveness that tops every city's list of objectives causes planning to give priority to growth at the expense of all other values, providing additional evidence to the critics who see it as serving developer interests at the expense of everyone else.

Still, despite the theoretical critique, practical difficulties of implementation, and inequitable outcomes so far, the progressive/leftist ideal of using planning to create a revitalized, cosmopolitan, just, and democratic city remains. Even while this vision seems forever chimerical, it continues as a latent ideal, bold in its scope but less rigid than its predecessors in regard to its general applicability. Frequently its content is assumed to be self-evident, but it is the measure against which practice is found wanting. Much of the critical planning literature, while attacking planning in practice, takes for granted that we know good and bad when we see it, freeing us from making elaborate arguments to justify our criteria. But using this critique implies that planning could do otherwise. In particular, the meaning of justice, which is not the only component of a good city but certainly one of the principal and perhaps most often transgressed elements, calls for sustained discussion.

Only recently has there been a move to develop principles of justice applicable to planning that at the same time recognizes the situational nature of ethical judgment:

> The just is not determined by an algorithm, particularly with respect to a situated activity such as planning. There is always scope for discretion, which in turn emphasizes that practical reasoning is essentially about judgment . . . It is about negotiating a path between the universal and the particular, leading to action.
>
> (Campbell 2006: 102)

This effort to describe the path between the universal and the particular is yet in a fledgling state. The purpose of this chapter is to move the discussion along. My focus is on justice and its relationship to democracy and diversity. These are elements of a theory of the good city, but a more encompassing discussion would also have to take into account a broader set of values, including desirable physical form, environmental protection, and authenticity. Such an analysis is beyond the scope of the discussion here.

THE CITY AS A SITE OF JUSTICE

The question of whether to pay particular attention to "the city" (or metropolitan area) needs discussion. Why not the region, the nation, the world? Paul Peterson (1981), in his book *City Limits*, contends that while city administrations can foster economic growth, they cannot engage in redistribution without precipitating capital flight, unemployment, and a decreasing tax base. Manuel Castells (1977) conversely asserts that cities are not the sources of production—that this is a regional function. If this is the case, and if production is key to the formation of economic interests, restricting analysis to cities or even metro areas is useless. In combination, Peterson and Castells could be interpreted as showing the irrelevance of the metropolitan level.

To be sure cities cannot be viewed in isolation; they are within networks of governmental institutions and capital flows. Robert Dahl, in a classic 1967 article, referred to the Chinese box problem of participation and power: at the level of the neighborhood, there is the greatest opportunity for democracy but the least amount of power; as we scale up the amount of decision-making power increases, but the potential of people to affect outcomes diminishes. The city level therefore is one layer in the hierarchy of governance. But the variation that exists among cities within the same country in relation to values like tolerance, quality of public services, availability of affordable housing, segregation/integration, points to a degree of autonomy. Justice is not achievable at the urban level without support from other levels, but discussion of urban programs requires a concept of justice relevant to what is within city government's power and in terms of the goals of urban movements (Fainstein and Hirst 1995; Purcell 2008: ch. 3). Moreover, there are particular policy areas in which municipalities have considerable discretion and thus the power to distribute benefits and cause harm; these include urban redevelopment, racial and ethnic relations, open space planning, and service delivery. Castells (1983), while minimizing cities' role in production, regards them as the locus of collective consumption—i.e., the place in which citizens can acquire collective goods that make up for deficiencies in the returns to their labor. Consequently he contends that urban social movements can potentially produce a municipal revolution even if they cannot achieve wider social transformation. According to this logic, then, urban movements do have some transformative potential despite being limited to achieving change only at the level in which they are operating.

In my attempt to develop criteria for a Just City, I first present an example of urban injustice then discuss the ways in which philosophers approach the value questions embedded within it. I then examine more generally the issues arising from various value criteria and their applicability to urban issues. The thoughts presented here represent the beginning of a larger project in which I attempt to develop an urban vision that can frame goals for urban development without being vulnerable to charges of moral absolutism.

AN EXAMPLE: THE BRONX TERMINAL MARKET

The story of a recent New York City planning decision with which I was peripherally involved as an advocate planner bears on the first three of the abovementioned policy areas (urban redevelopment, racial and ethnic relations, open space planning). It concerns the eviction of the wholesale food merchants at the Bronx Terminal Market, who were forced to leave their premises so that the market site could be turned over to a development firm. The market, which lies on city-owned land directly beneath the Major Deegan Expressway, opened in the 1920s and was renovated and reopened with considerable fanfare by Mayor Fiorello LaGuardia in 1935. Several of the last remaining firms could trace their origins back to those early days.

Figure 1.1 The Bronx Terminal Market

Reflecting the city's ethnic diversity, the merchants sold their exotic produce, meats, and canned goods primarily to bodegas, African food stores, and other specialized retailers. The city leased the market to a private firm, which collected rents and managed the facility. During the last few decades its neglect had resulted in decrepit structures, potholed roadways, inadequate services, grim interiors, and filthy surroundings. Those merchants who hung on to the end had suffered from the failure of the market's manager to maintain the property. Yet as late as 2005, 23 remaining wholesalers (down from an original peak of nearly 100) and their 400 employees were still generating hundreds of millions of dollars in sales.

In February 2006 the New York City Council approved the re-zoning of the market site from industrial to retail. The eight existing buildings, some of which had been listed on the National Register of Historic Places, would be torn down. The chairman of the Related Companies, which bought the market lease from the previous leaseholder, was a close friend of the city's deputy mayor for economic development. His firm intended to build a million-square-foot, suburban-style retail mall, to be called the Gateway Center at Bronx Terminal Market. It was to house a big-box retailer and a standard array of chain stores. Part of the project's financing depended on city and state subsidies, and the plans had to be approved by locally elected public officials, but no meaningful give and take ever took place between the merchants' association and city government. The Related Companies' glowing presentation of the project's putative benefits was never seriously challenged by any public official. Although the Bronx borough president, members of the community board, and city council members expressed sympathy for the plight of the merchants, who had endured decades of mistreatment, their sentiments did not move them to stand in the way of the juggernaut that was pushing the project. The treatment of the market comprises one more example of the priority given to economic development by city officials and the deployment of the utilitarian argument that decision rules should be based on the (alleged) greatest good of the greatest number.

The market merchants fought their displacement in court and before various city forums, including the local community board, the City Planning Commission, and the City Council. Sadly, however, the merchants lacked sufficient political influence to sway these officials either into willingness to integrate them into the Gateway project or to supply them with a suitable relocation site. By and large officials accepted the logic that the new mall represented necessary modernization and adaptation to the service economy.

Consultants to the market merchants proposed developing an integrated wholesale and retail market similar to the successful Pike Place Market in Seattle or New York's own Chelsea Market. Construction of a combined wholesale-retail facility would have differentiated the enterprise from cookie-cutter malls around the country, exploited its urban setting, and retained existing jobs. Use of a vacant city-owned site to the south of the market as well as a northern piece could have increased the area available for wholesale uses, but the city wished to reserve these areas for parkland as part of a swap for McComb's Dam Park, which it had designated as the location of a new Yankee Stadium (see note 8 below). The developer of the shopping mall was pre-selected without a public request for proposals or the opportunity for anyone to suggest other development strategies. Representatives of the city argued that since the developer bought the lease directly from the previous operator of the market, it was a purely private deal and therefore required no competitive bidding. Even though the city owned the land and leased the market buildings to a manager who had neglected them for years, the leaseholder was nevertheless allowed to sell his interest without constraints or effort by the city to buy back the lease itself. The city excluded affected residents and businesses from participating in planning for the area, limiting their input to reacting to the already formulated plan. The developer provided the neighborhood with at best minor concessions in the form of a community benefits agreement. Hundreds of well-paying jobs, almost all held by adult male immigrants, could be lost, replaced primarily by part-time, low-paid employment, and a once vital and viable business cluster was to be destroyed.[3]

A legal process was followed, as determined by judicial decisions, hearings were held, and the deal was carried out with the approval of the community board, the City Planning Commission, and the City Council. If evaluated by conformity to proper procedures, as prescribed by the city's land use review process, the outcome cannot be faulted. Decision-making occurred entirely at the local level, with no involvement of either the state or the federal government.[4] One could argue that structural forces (a changing economy, the need to compete) constrained decision-making. But the cases could have been decided differently and benefits could have been more equitably distributed.

The justification for the project primarily stems from its economic contribution rather than any physical improvements it will contribute to the area. One of the claims of the Gateway Center's developer is that the impoverished residents of the South Bronx crave the opportunity to shop at deep-discount stores. This is probably true, as New York has seen rising poverty and a declining median income since 1990. Residents are caught in a vicious circle: they cannot afford to patronize independent shopkeepers because their wages are so low, and their wages are so low because large corporations have been able to force down the general wage rate, justifying their stinginess as required by competition. The introduction of a big box store into this part of the Bronx will create more poorly paid employees who can only manage to patronize businesses that pay exploitative wages.

One's immediate reaction to this story is that injustice was done. The benefits of the project accrued to a wealthy developer and nationally owned chain stores. There was discrimination against small, independently owned businesses that were based in minority ethnic groups. Open space planning was conducted in a way intended to assist the New York Yankees rather than local residents. Can we demonstrate that these outcomes were unjust, and if so, by what criteria?

PROGRESSIVE VALUES

For the most part, empirical analysis, policy development, and theoretical formulation have proceeded on separate tracks.[5] Thus, my story of the Bronx Terminal Market resembles numerous other case studies in the urban literature that describe redevelopment projects and trace their outcomes to the power of the pro-growth coalition or the urban régime (Mollenkopf 1983; Logan and Molotch 1987; Fainstein and Fainstein 1986). They rarely, however, propose alternative policies. I can, however, list values that urbanists generally regard as goods and bads:

1. public space
   a. bad: lack of access, homogeneity
   b. good: heterogeneity
2. planning
   a. bad: rule of experts
   b. good: citizen participation

3. distribution of benefits
   a. bad: favors the already well-to-do
   b. good: redistributes to the worst-off
   c. ambiguous: assists the middle class
4. community
   a. bad: homogeneity
   b. good: recognition of "the other"; diversity[6]

While widely accepted, these criteria are rarely problematized or carefully justified.[7]

PHILOSOPHICAL APPROACHES

Philosophers, in contrast to urban scholars, spend their time developing and elaborating their ideas concerning justice. Their scrutiny, however, is rarely directed to urban issues, and their development of value criteria does not usually spell out appropriate urban policy.[8] Contemporary discussions of justice within philosophy nevertheless do concern themselves with questions that are also of central importance to urbanists and can therefore be extended to evaluating urban policy. Foremost are the questions of equality, democracy, and difference.

The work of John Rawls (1971) constitutes the usual foundation for discussions of justice and its relation to equality. As is well known, Rawls begins by positing an original position where individuals, behind a veil of ignorance, do not know their status in whatever society to which they will belong. Rawls' first principle is liberty and his second, subsidiary principle is "difference," by which he means equality. His argument is that free individuals, acting rationally, will choose a rough equality of primary goods so as to assure that they will not end up in an inferior position. Rawls has been so influential because, within a vocabulary acceptable to proponents of rational choice theory, he presents a logical argument that defends equality without resorting to natural law, theology, altruism, Marxist teleology, or a diagnosis of human nature.

Feminists, communitarians, and multiculturalists accuse Rawls of paying insufficient attention to other values besides primary goods, an obliviousness to social differences resulting from nonmaterial causes, and a failure to understand that society itself (i.e., community, interpersonal relations) is a good that is excluded by his emphasis on the individual. The question of whether Rawls' definition of primary goods can stretch to cover nonmaterial considerations does not concern us here but, as will be discussed below, issues of gender, cultural difference, and individualism do. Nevertheless, we can take away from Rawls for our purposes his justification of equality as a rational approach to organizing a "well-ordered society" or a well-ordered city.

Sen (1999) and Nussbaum's (2000) capabilities approach offers a further avenue for establishing values appropriate to the Just City. Capabilities do not describe how people actually function (i.e., end state) but rather what they have the opportunity to do. One need not exercise one's capabilities if one chooses not to (e.g., one can choose asceticism), but the opportunity must be available, including a consciousness of the value of these capabilities. According to this reasoning, each person must be treated as an end, and there is a threshold level of each capability beneath which human functioning is not possible. Thus, even if it could be demonstrated that the eviction of the Bronx merchants would produce the greatest good for the greatest number, the deprivation of their capability to earn a living could not be justified.

Nussbaum argues that capabilities cannot be traded off against each other. She lists, *inter alia*, life, health, bodily integrity, access to education, and control over one's environment (political and material) as necessary capabilities. Translated into a communal rather than individualistic ethic, the capabilities approach would protect urban residents from having to sacrifice quality of life for financial gain. Hence, for example, communities desperate for an economic base should not have to accept toxic waste sites because they lack any other form of productive enterprise. In contrast, conservative economists who support establishing market systems in pollution controls see such trade-offs as highly rational and to be desired.

The capabilities approach can be usefully applied to urban institutions and programs. In Sen's (1999: ch. 3) attack on utilitarianism he argues against the analysis typically employed by cost–benefit accounting as it is used to justify urban capital programs. These analyses typically exaggerate benefits and underestimate costs (Altshuler and Luberoff 2003; Flyvbjerg *et al.* 2003), rely on aggregates, and do not concern themselves with distributional outcomes. A more sensitive form of analysis asks who benefits and assesses what outputs each group in the population receives. Then, applying either the difference principle or the capabilities approach, we should opt for that alternative that benefits the least well-off. The definition of the least well-off, however, is subjective and is usually categorized according to social group affiliation. What we do know is that the group most lacking in political and financial power is least likely to prevail.

Nussbaum argues that false consciousness exists and that preferences are shaped, not simply there to be discovered. Thus, the welfare economics criterion of maximizing choice becomes undermined if people are deluded regarding the nature of a particular preference. Again to use the Bronx example, members of the City Planning Commission and the Bronx City Council delegation accepted unquestioningly the argument presented by the developers and city officials that residents of the Bronx would gain employment, amenities, and purchasing power through the construction of a shopping mall. They never were provided a developed conception of alternative forms of development or of a refurbished wholesale food market catering to ethnic cuisines; consequently their preferences were insufficiently informed. Or, to draw on an example of development policy prevalent in many American cities, we see one place after another investing large sums of public money in sports stadiums. Yet, numerous studies have shown that costs to the public fisc invariably outweigh benefits, that the surroundings of stadiums are dead spaces when no games are being played, and that the overall economic impact on the city is negligible. Politicians and much of the public, nonetheless, apparently accept the argument that cities benefit economically from the presence of major league teams.[9]

Likewise the Habermasian prescription of communicative rationality cannot operate in a situation where people are ignorant of their interests. Of philosophers Jürgen Habermas has probably had the most influence on the discipline of planning (see Forester 1993; Healey 2006). The ideal speech situation and concepts of deliberative democracy have particularly resonated within planning theory. Habermas's thought brings into play concepts of rationality, truth-telling, and democracy; its assumption is that through discourse, participants in decision-making will arrive at the best decision resulting from the force of the best argument. While offering criteria for evaluating the decision-making process, unlike Sen and Nussbaum, Habermas does not provide a metric for evaluating policy outcomes. As indicated by the Bronx Terminal Market and Yankee Stadium stories, the discourse surrounding planning usually involves false claims. A Habermasian approach would endorse exposing their falsity, but it would not point to how, in a field of power, this could be accomplished nor to what a just plan would involve. In the market case planners employed by the city unsurprisingly failed to challenge the shopping mall scheme or to present alternatives to it. Advocates for the merchants did so, but "speaking truth to power" had no effect.

Henri Lefebvre, like Iris Marion Young, is a philosopher who explicitly concerns himself with urbanism. His argument for "the right to the city" supports in particular the fight against the privatization of public space and the maintenance of heterogeneity within metropolitan areas (Lefebvre 1996; Mitchell 2003). As applied to the Bronx, it condemns the takeover of the Market by a speculative developer and the taking of public parks for a new Yankee Stadium. But, the "right to the city" lacks specificity, both in terms of what is included in that right and what is meant by the city. It is a vague concept that is more helpful as a rhetorical device than a policy-making instrument. At the same time it is useful to urban theorists because of its explicit concern with space, a variable excluded in most philosophical writing.

These philosophers then offer a route for considering planning actions and identifying their contributions to individual self-realization by providing criteria for evaluating planning methods and policy. The fairly glaring weakness of their arguments as practical tools is their lack of concern for the methods of achieving their ends, the absence of a formula for dealing with entrenched power, and their indifference to the costs and trade-offs that will be incurred by actually seeking to produce social justice. Nussbaum contends that it is unacceptable to trade capabilities against each other; that all must be achieved. This, however, may not be possible. Unlike Marx, who criticized the utopians for their failure to identify a means to achieving their ends, contemporary political philosophers apparently feel that implementation is someone else's concern. But planners, policy makers, and political activists cannot wipe out history and act as if they start from scratch—they have to be contextualists. While utopian ideals provide goals toward which to aspire and inspiration by which to mobilize a constituency, they do not offer a strategy for transition within given historical circumstances. As Marx reminded us, people make their own history, but not under circumstances of their own making. Original positions, desired capabilities, ideal speech situations, and rights to the city seem remote from the actualities of the Bronx.

PRACTICALITIES

All these endeavors at providing a normative framework need to be examined in relation to practical realities of régime formation, social exclusion, and the bases of conflict, and they ought to take into account the variation among places. Each philosophical line of attack presents serious issues. Rawls justifies the value of equality by arguing that, in the original position, behind a veil of ignorance, each individual would opt for it. We are, however, never in the original position (as is argued by communitarian critics). Rawls' "difference principle" (ironically) evades questions of difference based on disability or multiculturalism (see Young 1990; Nussbaum 2006). Equality of primary goods does not compensate for physical incapacity or disrespect.[10] The Market merchants who sold goods to bodega owners did not receive the same deference as the owners of the New York Yankees. Not only were they treated unjustly in the sense that they did not receive accommodation that would allow them to continue their businesses in a shared facility, but their outrage and alternative proposals were treated with indifference and even contempt.

Nor does the difference principle deal with the loss of liberty derived from obligations arising from family responsibilities and the limitations they place on freedom.[11] Thus, Rawls has been criticized as overly materialistic (Hirschman 1989; Jaggar 1983; Young 1990: 16; Nussbaum 2000: ch. 1). Within rational choice theory it is always possible to compensate an individual for loss. In law, for instance, this outlook is codified in the payment of damages—you lose an arm in an industrial accident, and you receive X amount of dollars according to a Workman's Compensation schedule. But we know that an arm is not really equivalent to any amount of money. Likewise one cannot escape the obligations of parenthood by paying a babysitter. The creation of job opportunities through growth promotion without accompanying social and transportation services causes poor people to face unenviable choices between caring for family members, incurring difficult commutes, and gaining employment. Planning involves not only consideration of the financial impacts of a particular policy but also its effects on people's well-being: Do its benefits make up for the humiliations of a demeaning job? Does replacement of lost park land by an equal amount of other, less accessible space make up for the loss? Do enhanced revenues and improved social services justify breathing polluted air?

Starting with the individual leads to a discussion of equality among individuals rather than of social relationships among collectivities. Much of philosophical discussion in relation to justice thus revolves around the question of the desirability of equality based on primary goods—for example, whether or not handicapped individuals should receive the same amount of primary goods as everyone else or whether they should receive additional, compensatory benefits (Anderson 1999; Nussbaum 2006). A more sociological discussion would employ the term equity instead and concern itself with redressing disadvantage as its affects groups (see Campbell 2006, pp. 94–95).

Equity leads us to include a broad range of considerations that concern us as planners—for example, the impacts of environmentally degrading facilities on different social groups, or who has access to public space and for what purposes public space can be used. It points to the results of public policies rather than to simply the analysis of starting points. By examining outcomes in relation to groups we avoid utilitarian cost–benefit analyses that focus on aggregates of individuals and we also have a better handle on power relations and social structures.

Failure to acknowledge the coherence of collectivities and their structural relationships to each other evades a fundamental social issue of redistribution—how to avoid a burden that the better-off refuse to accept. Serious redistribution inevitably provokes conflict that may be so severe as to destroy social peace. The practical issue of moving toward greater equality then is to devise ways of mitigating the adverse effect on those forced to give up a great deal.

The starting point of individual liberty also avoids questions that bear on the character of collective goods—e.g., a high-quality built environment—if they are not necessary for the development of individual capabilities or remedying inequality. A recent debate on Chicago that appeared on the urban sociology listserv concerned whether the creation of a lively city with attractive amenities has widespread benefits or whether it is only pertinent to *bourgeois* consumers while low-income groups continue to suffer from social exclusion (Gilderbloom 2006). As phrased in this particular discussion, there seemed to be an underlying assumption that low-income people do not care for amenities. In other words, it is implied that city beautification matters only to urban élites and that working class people care only for material benefits. Once, when I was teaching in New Brunswick, NJ, I asked a local minister, who was lecturing to my class, whether his congregation, which mainly resided in public housing, resented the transformation of downtown by brick sidewalks and street furniture. Did he feel that their space was being taken away from them for the benefit of young urban professionals? "Are you serious?" he replied. "Do you think my people don't like to be somewhere that looks nice?" Reaction against exclusionary practices seems to have devolved into regarding an association between low-income people and ugly surroundings as desirable.[12] The right to the city, however, refers to more than mere inclusion—it encompasses access to an appealing and sustaining city and the development of an urbane environment and also to participation in shaping that environment:

> Lefebvre's vision isn't only for a more user-centered design of concrete space . . . The right to appropriation can be conceived not just as the right to be physically present in existing urban space, but the right to a city that fully meets, above all other considerations, the needs of inhabitants . . . Appropriation in that larger sense would mean a right to a city where workers could make a short commute to work . . . and come home to affordable comfortable housing. It would allow child-care-givers to choose from several nearby parks . . . It would mean shoppers visiting a nearby grocery store that offered high-quality, reasonably priced food. It would mean a city without racial and class segregation that reinforced social inequalities. Certainly appropriation demands the right to be present in space, but it also requires the production of spaces that actively foster a dignified and meaningful life.
>
> (Purcell 2008: 95)

Philosophers have had to take account of the post-modernist/post-structuralist emphasis on the situatedness of the speaker and its assault on the existence of a unitary ethic. Even though economic disadvantage coincides with ethnic and religious difference, material equality cannot by itself overcome issues arising from lack of what philosophers term "recognition"; hence the debate over whether recognition subsumes redistribution or whether, as Fraser contends, both must be analyzed separately despite their real-life entanglement (Fraser and Honneth 2003). Those like Nussbaum (1999: ch. 1) who seek to retain a universalistic ethic maintain that, despite cultural difference, there is a broad common value structure that can stretch to embrace tolerance and difference itself. Rawls (1971, 2001) recognizes the invidiousness of difference and comments that even a redistributional welfare state fails to produce justice, because it concentrates the control of productive resources in one group and thereby produces a stigmatized disadvantaged class of aid recipients. Current developments in the welfare states of Northern Europe, where income support for unemployed citizens from mainly immigrant backgrounds does not succeed in quelling their anger at their situation, validates Rawls' argument.

Rawls opts for either a "property-owning democracy" (i.e., widely distributed ownership of productive assets) or liberal socialism as the basis for a "well-ordered society." The question comes up again: how do we get there? What arguments can make people accept redistribution if they already know that they are in an advantaged position? It cannot be simply how it would feel to be in the other person's place if we already know we are not. This is a particularly acute problem if those who are advantaged identify the disadvantaged as "other" in terms of ethnicity, religion, or color. There needs to be an argument based on collective good—social rationality—rather than simply individual rationality, even though it need not be a strictly utilitarian one. And, in practical terms, it must be backed by the force of a social movement, a political party, or a supportive élite.

Is it feasible to move toward this desired state (of property-owning democracy or liberal socialism) at the urban level? It would be easier if more goods that are controlled at the local level were universally publicly provided. For example, in London entrance fees for public museums were scrapped after the national Labour government changed the tax laws, while the local Labour authority started taxing cars that entered central London; we take the public library for granted—why should not other entertainment/and educational providers also offer free or very inexpensive services? (New York City is going in the opposite direction and has introduced fees into formerly free recreation centers and during the 1975 fiscal crisis ended free tuition at the City University.) In London during the days of the Greater London Council, fares were significantly lowered on public transit. Such a measure is probably the single most efficacious redistributional strategy available to local government. Moreover, the more that the whole society has a stake in collective goods as a consequence of public provision, the more reform ("voice") rather than exit will operate to maintain their quality for everyone (Hirschman 1970).

Under the property-owning democracy formulation, home ownership becomes a desirable goal and the "taking" of private homes for economic development purposes is wrong.[13] Widespread home ownership makes available greater use values in housing for people, but it has the drawback of introducing a speculative financial element into the enjoyment of shelter as well as being inappropriate for households that do not have the resources to cope with system breakdowns or even routine maintenance.[14] We can, however, look to the examples of Amsterdam and Stockholm, where public ownership of land does not inhibit private development of structures but retains increases in land value for the public and makes renting a good choice for many. Even in New York City the World Trade Center and Battery Park City sites are publicly owned and the owners of structures pay land rent (although this situation did not save the Bronx Terminal Market merchants).

GROWTH, EQUITY, AND DIVERSITY

The most politicized urban issues usually revolve around a conflict between the goals of growth and equity. There is a tendency among critics of redevelopment programs to regard growth as a negative aim—ecologically damaging, with its benefits going to the already affluent. But the benefits of growth would be more widely distributed if ownership were less concentrated, as in the property-owning democracy model.

I.M. Young starts with social institutions rather than individuals in her analyses. She deals especially with the relationships between diversity and equality, distinctive cultural practice and social exchange, difference and integration. In *Inclusion and Democracy* (2000) she takes the position that more democracy will produce more equality, but she considers that the concept of deliberative democracy, as it is usually framed, is impractical in mass societies—it is too time-consuming and requires face-to-face interaction. It is not clear, however, as to why her approach, which accepts conflict and irresolution, is more practical in terms of arriving at desirable substantive outcomes. And, in fact, it differs little from Habermas', who also speaks of decentered democracy.

Young (2000) argues for "differentiated solidarity" rather than integration—i.e., geographical groupings with fuzzy borders. Here she identifies a realistic approach to the issue of multiculturalism, which is somewhat at odds with Lefebvre's right to the city and the criterion that urban spaces should be highly heterogeneous. Efforts to force residential integration have too frequently been counter-productive—not just in terms of backlash but also in depriving groups of mechanisms for mutual support. Residential differentiation need not necessarily imply lack of mixing elsewhere—in public spaces, at work, in recreational areas, and at school. There is criticism of New York's Battery Park City as being a virtual gated community (Kohn 2004), yet anyone can in fact gain access to its open space and indoor Winter Garden (unlike the Bronx Zoo, which charges a steep admission fee despite its location in the heart of New York's poorest borough). Every public space need not be used by a full range of inhabitants, but should also not keep people out. Cities can be diverse and tolerant in macro without each sub-area encompassing a multi-ethnic, multi-class citizenry. A far greater danger than public spaces with iconography that seems forbidding to some is homogeneous municipalities of rich, poor, and middle on the periphery, not separation within the city itself, as long as internal boundaries are porous.

Conservative values of order and efficiency may clash with those of equality and diversity. The left dismisses the former as supportive of privilege and legitimated through propaganda (Sennett 1970; Foucault 2003). But these are values that enjoy wide popular support and are essential to the functioning of society. Hobbes' argument that maintenance of personal safety is the first duty of the sovereign cannot be dismissed as simply a rationalization for authoritarian rule. We need to find out how to interpret these conservative values in humanitarian ways whereby they do not suppress dissent, produce sterile environments or only benefit the rich, but we cannot simply disregard them.

In past work I have portrayed Amsterdam as providing an actual model of social justice. Recently its success has been questioned as result of decline in the Dutch welfare state, ethnic friction, and the tightening of rules for immigration (see Uitermark 2003; Kramer 2006; Dias and Beaumont 2007). Still, it continues to support a great deal of social and political equality, diversity and integration, planning and economic growth. The Amsterdam case implies democratic procedures and just actions flow from situations where rough social justice already exists. While the criteria of social justice may

Figure 1.2 Amsterdam

transcend particular social contexts, its implementation requires that elements of realization be already present. Achievement of the just is a circular process, whereby the preexistence of equity begets sentiments in its favor, democratic habits produce popular participation, and diversity increases tolerance. The sobering lesson of present-day Amsterdam, however, shows that even virtuous circles can be destabilized and that disruption, as occurred with the assassination of Theo Van Gogh, can precipitate a chain of events that easily breeds intolerance and fear of difference (Baruma 2006). Moreover, victims can also be victimizers.

PROCESS AND OUTCOME

We can now return to the values presented earlier in the chapter as taken for granted by progressive urbanists. The various examples discussed here indicate the way that process and outcome are tangled with each other and how particular values, depending on their context and interpretation, can cut against each other. Most troublesome is the tension between democracy and equality. Consider the case of the Bronx Terminal Market. Formal democratic processes were followed. Unlike the Yankee Stadium instance, the plan for the shopping mall was considered by a number of public bodies, including the local community board. There were also several meetings involving the merchants and their advocates, public officials, and the developer that conformed to a deliberative model. No consensus transpired, however, because the imbalance of power was too great. A viewpoint demanding substantive equity would have provided a place for the merchants in the restructured Bronx, but such an outcome could not occur without initially greater equality among the participants. If we apply the capabilities approach, we see that a particular group was denied their ability to function and also denied recognition.

When we think about planning for cities, therefore, we must realize that substance and procedure are inseparable. Open processes do not necessarily produce just outcomes. Proceeding from a situation lacking in supportive values to a more enlightened state presents baffling strategic problems, because mobilizing a force sufficient to overcome barriers to change demands a messianism that contravenes undistorted speech and can provoke fierce reaction. But, just as substance and procedure must be contemplated simultaneously, so must desirable end states and the forces to achieve them. If Amsterdam presents a rough image of a desirable urban model, strategies and normative emphases will differ in respect to reaching that goal depending on starting point.

In the United States distributional issues are especially salient because social citizenship has not yet been won (Marshall 1965). Justice requires dampening of sentiments based on group identity, greater commitment to common ends, and identification of institutions and policies that offer broadly appealing benefits. As it is, in the U.S. no broad-based media exist to communicate alternative approaches to questions raised by urban economic development, metropolitan inequalities, and environmental preservation. The inherently divisive character of identity politics cuts against the building of such institutions and therefore is largely self-defeating. For redistribution to proceed, recognition is a prerequisite, but that recognition needs to involve shared commitments not rivalry.

The historically most effective approach to urban political transformation in the United States used group identity to bolster unity toward greater ends than symbolic recognition. During the 1960s successful movements in the U.S. were based in groups that shared racial, territorial, and client statuses (Fainstein and Fainstein 1974). This neighborhood base, with community control as its objective, has, however, lost its force as a consequence of immigration, gentrification, and racial integration of the civil service (Fainstein and Fainstein 1996). In the new century, effectiveness probably means organizing around work status when it overlaps with racial, immigrant or gender situation (living wage movements). Whereas the urban social movements of the past centered on collective consumption (Castells 1977), future movements need to address the organization of work as well as concerning themselves with the consumption issues of new types of workers.[15] The changing nature of work calls for unions of temporary workers, household workers, and the self-employed rather than traditional organizing around the workplace.[16] Such unions would have to emphasize their service role: job training and placement; establishment of benefit pools and portability of benefits; provision of legal services; credit unions, and mortgage assistance. This also means continued organizing around affordable housing, but to be successful such programs would have to recognize the housing needs of the middle class, not simply call for assistance to the poorest. Narrowly targeted policies, however efficient, lack a sufficient constituency and seem unjust to those not benefiting.

Citizen participation's importance also varies with context. In most European cities, there is no absolute material need on the American scale. Especially in France and Germany, the plea for citizen participation, negotiation, and a less authoritative government makes sense. Within this context a more transactional approach represents reform. In the U.S., where most cities are dominated ideologically as well as politically by business-led régimes and homeowner groups rather than public bureaucracies, individual citizen participation will not provide a path to social transformation even though it can block destructive projects. Urban citizen participation mainly involves participants demanding marginal changes in the *status quo* or benefits that respond to their narrowly defined interests.

The movement toward a normative vision of the city requires the development of counter-institutions capable of reframing issues in broad terms and of mobilizing organizational and financial resources to fight for their aims. Castells (2000: 390) doubts the usefulness of abstract conceptions of justice; he fears that visionary projects lead only to grief. But there is a need to persuade people to transcend their own narrow self-interest and realize that gains can be had from the collective enterprise. Such a mobilization depends on a widely felt sense of justice and sufficient threat from the bottom to induce redistribution as a rational response. Enough of the upper social strata need to accept a moral code such that they do not resist, and will even support, redistributional measures.

Thus, when thinking about just cities, we must think simultaneously about means and ends, social movement strategies and goals as well as appropriate public policy. In the past moves toward progressive ends have arisen from both popular demands and insulated bureaucracies (Flora and Heidenheimer 1981). We cannot know, *ex ante*, what will be the most fruitful source of change, but by continuing to converse about justice, we can make it central to the activity of planning. The very act of naming has power. If we constantly reiterate the call for a Just City (as conservative forces forever refer to economic development and the Congress for the New Urbanism talk about smart growth and stopping sprawl), we change popular discourse and enlarge the boundaries of action. Communicative theorists are right in emphasizing the importance of words, but for justice to prevail, it is imperative that the content of speech include demands for recognition and just distribution. Changing the dialogue, so that demands for equity are no longer marginalized, would constitute a first step toward reversing the current tendency that excludes social justice from the aims of urban policy.

NOTES

1 A shorter, somewhat different version of this chapter appears in *Harvard Design Magazine*, 27, Fall 2007/Winter 2008, pp. 70–76. Thanks to Peter Marcuse, Norman Fainstein, Sharon Meagher, and the editors of this volume for their helpful comments.

2 New York's PlaNYC and plans for London's Thames Gateway are unusual in the scope of their ambition.

3 The mall will present long, blank exterior walls, offering only a few corridors into the surrounding neighborhood. In addition, the big-box store is strongly opposed by local unions because of the employment practices of this type of merchandiser. Much of the site will be dedicated to parking decks. The architects' renderings of the center show urbane visitors sipping cappuccino at outdoor sidewalk cafés flanking the mall. Presumably these boulevardiers would ignore the noise, soot, exhaust, and bird droppings drifting down from the highway underside immediately above them.

4 Some state tax benefits will accrue to the developer as a consequence of the site's location in an Empire State Development Zone. The state, however, was not an active participant in the decision-making process.

5 Some urbanists have started to formulate ethical theories (*inter alia*, Andrew Sayer and Michael Storper, John Forester, Richard Sennett, Heather Campbell), and some political theorists and philosophers have directly or indirectly concerned themselves with urban issues (*inter alia*, Frank Fischer, Martha Nussbaum, Iris Marion Young).

6 Other goods and bads include: *quality of the built environment*—bad: inauthenticity, conformist architecture/good: historical accuracy, cutting-edge architecture; *social control*—bad: order/domination/good: resistance/conflict; *housing*—bad: luxury dwellings/good: affordable units; *mega-projects*—bad: large, top-down planned/good: popular, incremental, preservation; *social services*—bad: privatization, individualization/good: collective consumption; *economic development* —bad: entrepreneurial state/good: small business, cooperatives; *environment*—bad: *laissez-faire*/good: regulation, green development.

7 The work of David Harvey (1992, 2003), particularly his discussion of Paris and his article analyzing the conflict over Tompkins Square Park, is an important exception to this generalization.

8 Iris Marion Young is unusual in having concerned herself specifically with urban questions, perhaps as a consequence of teaching for years within a planning program at the University of Pittsburgh.

9 See Rosentraub (1999). New York City and State have recently committed themselves to invest $362 million in a new Yankee Stadium, even though the rebuilt stadium will occupy what had been public park land. The justification was its contribution to the economic development of the Bronx even though the present stadium operates at capacity and there is no danger that the team will leave New York. The benefit of the new stadium is entirely to the team, which will gain from more extensive skyboxes and more space for ancillary, money-making facilities. Despite local protest over the loss of parks, there has been no major opposition to this expenditure of public funds in the face of much more obvious needs in this impoverished borough. The city is supplying replacement park land, but it is broken up and much less accessible than what is being taken away.

10 Rawls (1971: 440) does state that self-respect is "perhaps the most important primary good." One of the two components of self-respect is "finding our person and deeds appreciated and confirmed by others."

11 Rawls seems to regard the family as a single unit and does not examine what would constitute justice within it.

12 Within the U.S. context, where controls on gentrification are minimal, this association is understandable. Indeed once downtown New Brunswick became desirable, its public housing was demolished and replaced by condominiums, with most of the inhabitants pushed to the city's outskirts.

13 The U.S. Supreme Court ruled in the case of *Kelo v. New London* (June 23, 2005) that municipalities could use the power of eminent domain to take private property in order to turn it over to another private party in furtherance of the aim of economic development.

14 The US government's emphasis on home ownership and its subsidization through the tax code are generally repudiated by progressives. But ownership increases people's feelings of autonomy and protects them from exploitation by landlords. Limits on the size and use of the tax deduction, however, may be justifiable.

15 Ira Katznelson (1981) has called attention to the enduring split between home and work as a cause of American workers' political failures. Organizing successes by the Los Angeles Alliance for a New Economy, which has built on community–labor ties, however, points to the potential of strengthening labor's community base. (See http://www.laane.org for a summary of its initiatives.)

16 The first National Domestic Workers Congress took place in New York on the weekend of January 6, 2008 (Buckley and Correal 2008). Most of the attendees were women representing locally organized groups in a number of cities. They aimed to build alliances and develop strategies to demand benefits including paid vacations and holidays, cost-of-living wage increases, health insurance and advance notice of termination and called for a proposed New York State Domestic Workers' Bill of Rights which, if passed, would be the first in the nation. The nature of domestic work means that organizing has to be on a community rather than a workplace basis.

References

Altshuler, A. (1965) *The City Planning Process: A Political Analysis*, Ithaca, N.Y., Cornell University Press.

Altshuler, A. and Luberoff, D. (1965) *The City Planning Process*, Ithaca, N.Y., Cornell University Press.

—— and —— (2003) *Mega-projects*, Washington, DC: Brookings Institution.

Anderson, E. (1999) "What is the point of equality?" *Ethics*, 109: 287–337.

Anderson, M. (1964) *The Federal Bulldozer; A Critical Analysis of Urban Renewal, 1949–1962*, Cambridge: MIT Press.

Baruma, I. (2006) *Murder in Amsterdam: The Death of Theo Van Gogh and the Limits of Tolerance*, New York: Penguin.

Buckley, C. and Correal, A. (2008) "Domestic workers organize to end an 'atmosphere of violence' on the job," *New York Times*, June 8.

Campbell, H. (2006) "Just planning: the art of situated ethical judgment," *Journal of Planning Education and Research*, 26: 92–106.

Castells, M. (1977) *The Urban Question*, Cambridge: MIT Press.

—— (1983) *The City and the Grassroots*, Berkeley: University of California Press.

—— (2000) *End of Millennium.* 2nd edn, Oxford: Blackwell.

Dahl, R.A. (1967) "The city in the future of democracy," *American Political Science Review*, 61: 953–970.

Davidoff, P. and Reiner, T. (1962) "A choice theory of planning," *Journal of the American Institute of Planners*, 28: 103–115.

Dias, C. and Beaumont, J. (2007) "Beyond the egalitarian city," paper presented at Committee on Urban and Regional Research (RC-21) of the International Sociological Association Conference on Urban Justice and Sustainability, Vancouver, Canada, August.

Fainstein, N. and Fainstein, S.S. (1986) "Regime strategies, communal resistance, and economic forces" in S.S. Fainstein, N.I. Fainstein, R.C. Hill, D. Judd, and M.P. Smith, *Restructuring the City*, rev. edn, New York: Longman, pp. 245–282.

—— and —— (1996) "Urban regimes and black citizens: the economic and social impacts of black political incorporation in US cities," *International Journal of Urban and Regional Research*, 20: 22–37.

Fainstein, N. and Fainstein, N.I. (1974) *Urban Political Movements*, Englewood Cliffs, NJ: Prentice-Hall.

Fainstein, N. and Hirst, C. (1995) "Urban Social Movements," in D. Judge, G. Stoker, and H. Wolman (eds) *Theories of Urban Politics*. London: Sage, pp. 181–204.

Flora, P. and Heidenheimer, A.J. (1981) *Development of Welfare States in Europe and America*, rev. edn, New Brunswick, NJ: Transaction Books.

Flyvbjerg, B., Bruzelius, N., and Rothengatter, W. (2003) *Megaprojects and Risk*, New York: Cambridge University Press.

Forester, J. (1993) *Critical Theory, Public Policy, and Planning Practice: Toward a Critical Pragmatism*, Albany, NY: State University of New York Press.

Foucault, M. (2003) *The Essential Foucault*, ed. P. Rabinow and N. S. Rose, New York: New Press.

Fraser, N. and Honneth, A. (2003) *Redistribution or Recognition?* trans. J. Golb, J. Ingram, and C. Wilke, London: Verso.

Gans, H.J. (1968) *People and Plans*. New York: Basic Books.

Gilderbloom, J. (2006) "What is success? Chicago and around the world—my own meditation on the subject," [Online Listserv] communication to Comurb listserv. Available at: https://email.rutgers.edu/mailman/listinfo/comurb_r21 [accessed April 2, 2006].

Hall, P.G. (2002) *Cities of Tomorrow*, Oxford: Blackwell.

Harvey, D. (1978) 'Planning the ideology of planning', in R. Burchell and G. Sternlieb (eds), *Planning Theory in the 1980s*, New Brunswick, NJ: Rutgers University Center for Urban Policy Research, pp. 213–234.

—— (1992) "Social justice, postmodernism and the city," *International Journal of Urban and Regional Research*, 16: 588–601.

—— (2003) *Paris, Capital of Modernity*, New York: Routledge.

Hayek, F.A. von (1944) *The Road to Serfdom*, Chicago, University of Chicago Press.

Healey, P. (2006) *Collaborative Planning: Shaping Places in Fragmented Societies*, New York: Palgrave Macmillan.

Hirschman, A.O. (1970) *Exit, Voice, and Loyalty*, Cambridge: Harvard University Press.

Hirschmann, N.J. (1989) "Freedom, recognition, and obligation: a feminist approach to political theory," *American Political Science Review*, 83: 1227–1244.

Jacobs, J. (1961) *The Death and Life of Great American Cities*, New York: Vintage.

Jaggar, A.M. (1983) *Feminist politics and human nature*, Totowa, N.J. Rowman & Allanheld.

Katznelson, I. (1981) *City Trenches*, Chicago: University of Chicago Press.

Klosterman, R. (1985) "Arguments for and against planning," *Town Planning Review*, 56: 5–20.

Kohn, M. (2004) *Brave New Neighborhoods: the Privatization of Public Space*, New York: Routledge.

Kramer, J. (2006) "Letter from Europe. The Dutch model: Holland faces its radical Muslims," *The New Yorker*, April 3: 60–67.

Lefebvre, H. (1996) *Writings on Cities*, ed. and trans. E. Kofman and E. Lebas, Oxford: Blackwell.

Lindblom, C.E. (1959) "The science of 'muddling through,'" *Public Administration Review* 19(2): 75–88; reprinted in S. Campbell and S.S. Fainstein (eds) (2003), *Readings in Planning Theory*, 2nd edn., Oxford: Blackwell, pp.196–209.

Logan, J.R. and Molotch, H. (1987) *Urban Fortunes*, Berkeley: University of California Press.

Marshall, T.H. (1965) *Class, Citizenship, and Social Development*, New York: Anchor.

Mitchell, D. (2003) *The Right to the City: Social Justice and the Fight for Public Space*, New York: Guilford Press.

Mollenkopf, J.H. (1983) *The Contested City*, Princeton: Princeton University Press.

Nussbaum, M.C. (1999) *Sex and Social Justice*, Oxford: Oxford University Press.

—— (2000) *Women and Human Development: The Capabilities Approach*, Cambridge: Cambridge University Press.

—— (2006) *Frontiers of Justice*, Cambridge: Harvard University Press.

Peterson, P. (1981) *City Limits*, Chicago: University of Chicago Press.

Purcell, M. (2008) *Recapturing Democracy*, New York: Routledge.

Rawls, J. (1971) *A Theory of Justice*, Cambridge: Harvard University Press.

—— (2001) *Justice as Fairness*, in Erin Kelly (ed.), Cambridge: Harvard University Press.

Rosentraub, M. (1999) *Major League Losers: The Real Cost of Sports and Who's Paying for It*, revised edn, New York: Basic Books.

Scott, J.C. (1998) *Seeing like a State*, New Haven: Yale University Press.

Sen, A. (1999) *Development as Freedom*, New York: Anchor.

Sennett, R. (1970) *The Uses of Disorder*, New York: Knopf.

Thomas, J.M. (1996) "Educating planners: unified diversity for social action," *Journal of Planning Education and Research*, 15: 171–82.

Uitermark, J. (2003) "'Social mixing' and the management of disadvantaged neighbourhoods: the Dutch policy of urban restructuring revisited," *Urban Studies*, 40: 531–549.

Yiftachel, O. (1998) "Planning and social control: exploring the dark side," *Journal of Planning Literature*, 12: 395–406.

Young, I.M. (1990) *Justice and the Politics of Difference*, Princeton: Princeton University Press.

—— (2000) *Inclusion and Democracy*, New York: Oxford University Press.

2 The right to the Just City

David Harvey with Cuz Potter

Susan Fainstein's conception of the Just City is at heart an attempt to philosophically define a harmonious and just urban form. While identifying and defining the city of one's heart's desire constitutes a vital moment in working towards a better future, it can only be useful to the extent that it grounds itself within the concrete, contemporary social context and engages in the effort to reshape it. This context is a three-decade long restoration of élite class power through the roll out of neoliberal policies based on a discourse of individual rights and freedoms. Counteracting this consolidation of class power requires struggle to redefine rights and freedoms and reconfigure the social processes upon which they rest. Urban transformation plays a central role in this struggle, and thus we fight for the right to a Just City.

To date, Fainstein's extensive exploration of the potential urban manifestations of social justice has provided us with only a hazy outline of what might constitute the "Just City." I do not think this is an accident. For what is social justice? Thrasymachus in Plato's Republic argues that "each form of government enacts the laws with a view to its own advantage" so that "the just is the same everywhere, the advantage of the stronger." Plato rejected this cynical view in favor of justice as a specifiable ideal (Plato 1955: 66). A plethora of ideal formulations now exist from which we can choose. We could be egalitarian; utilitarian in the manner of Bentham (the greatest good of the greatest number); contractual in the manner of Rousseau (with his ideals of inalienable rights) or John Rawls; cosmopolitan in the manner of Kant (a wrong to one is a wrong to all); or just plain Hobbesian, insisting that the state (Leviathan) impose justice upon reckless private interests to prevent social life from being nasty, brutish, and short. Some writers, disturbed by the deracinated universalism of such theories of justice, even argue for local ideals of justice, sensitive to cultural and geographical differences (Walzer 1983). We stare frustratedly in the mirror asking: "which is the most just theory of justice of all?" In practice, we worry that Thrasymachus might have been right: justice is simply whatever the ruling class wants it to be. And when we look at the history of jurisprudence and of judicial decisions and how these have evolved in relation to political power, it is very hard to deny that ideals of justice and practices of political power have marched along very much hand in hand. Foucault, for one, has plausibly questioned again and again the class neutrality of any discourse on rights, let alone of any system of adjudication through the courts (Foucault and Gordon 1980).

Yet we cannot do without the concept of justice for the simple reason that the sense of injustice has historically been one of the most potent seedbeds of all to animate the quest for social change. The idea of justice in alliance with notions of rights has not only been a powerful provocateur in political movements but the object of an immense effort of articulation. The problem is, therefore, not to relativize ideals of social justice and of rights but to contextualize them. When we do that we see that certain dominant social processes throw up and rest upon certain conceptions of rights and of social justice. To challenge those particular rights is to challenge the social process in which they inhere. Conversely, it proves impossible to wean society away from some dominant social process (such as that of capital accumulation through market exchange) to another (such as political democracy and collective action) without simultaneously shifting allegiance from one dominant conception of rights and of social justice to another. The difficulty with all idealist specifications of rights and of justice, including Fainstein's Just City, is that they hide this connection. Only when they come to earth in relation to some social process do they find social meaning.

This problem is highlighted in John Rawls' theory of justice as fairness, from whom Fainstein (this volume) has drawn the justification of equality as a rational approach to organizing a well-ordered city. Rawls seeks a neutral standpoint from which to specify a universal conception of justice. He constructs a "veil of ignorance" concerning the position we might occupy in the social order and asks how we would specify a just distribution in the light of that ignorance. But he cannot presume total ignorance since nothing whatsoever could then be said. He therefore assumes that we know the general laws of human psychology and of economic behavior, that we are familiar with the dominant social processes through which the social order is reproduced. But these are not universal truths. In what historical time and geographical place do I locate myself? To what school of economic or psychological thought do I attach myself? If I can choose between classical political economy, marginalist economics (with its thesis that fairness is given by the marginal rates of return on scarcity of land, labor, and capital), or some version of deep ecology, Marxian, or feminist theory, then outcomes will plainly be quite different. Hardly surprisingly, Rawls' system ends up largely confirming notions of rights inherent in the market and in *bourgeois* society, even as it concedes that there is no real way to adjudicate between socialist and capitalist ways of doing things (Rawls 1971).

We therefore need to shift attention away from consideration of abstract universals towards the relation of concepts of rights and of justice to social processes. Consider, for example, the dominant social processes at work within our own world. They cluster around two dominant logics of power: that of the territorial state and that of capital (Harvey 2003, 2006). These two logics of power are often in tension if not outright opposition to each other at the same time as they must in some way fulfill and support each other lest social reproduction dissolve into total anarchy and nihilism.

Consider, first, the territorial logic of state powers. However much we might wish rights to be universal—as the declaration of the universal rights of man first envisaged—it requires the protection of the state apparatus to enforce those rights. But if political power is not willing, then notions of rights remain empty. Rights in this instance are fundamentally derivative of and conditional upon citizenship and territorialized power (primarily but not uniquely expressed as state power). The territoriality of jurisdiction then becomes an issue. This cuts both ways. Difficult questions now arise because of stateless persons, migrants without papers, illegal immigrants, and the like. Who is or is not a "citizen" becomes a serious issue defining principles of inclusion and exclusion within the territorial specification of the national or local state (in which jurisdiction, for example, can I cast my vote?). How the state exercises sovereignty is itself a huge issue and as has been asserted in numerous writings in recent years, there are limits placed on that sovereignty (even at the national as well as at the local level) by the rules that govern the circulation and accumulation of capital across the globe. Nevertheless, the nation state, with its monopoly over legitimate forms of violence, can in Hobbesian fashion define its own bundle of rights and interpretations of rights and be only loosely bound by international conventions.

Urban rights and justice are therefore mediated by the spatial organization of political powers. Patterns of urban administration, policing, and regulation are all embedded in a system of governance that allows for the playing out of multiple interests in the murky corridors of urban politics and through the labyrinthine channels of urban bureaucracy and administration. Certain rights are coded within these systems but others are simply denied, or, more likely, rendered so opaque by bureaucratic fudging as to be meaningless. Planning powers (this zoned for commerce, that condemned for insalubrity), edicts to regulate behavior ("no loitering here"), surveillance (video cameras on every corner), lop-sided service provision (clean streets here, garbage dumps there), and the desperate attempt to impose order, suppress crime and conflict, and bring regularity to daily life in the city—these are everywhere in evidence. Urban citizenship (the rights of immigrants, transients, and strangers to participate in local politics) is an even murkier concept than that of the state, since it so often depends on residence and domicile in a social world that is now more than ever constructed on principles of motion. State powers are invariably obsessed with maintaining order and erasing difference when both disorder and difference are fundamental to the creativity of urban life. The contemporary so-called "War on Terror" is dangerously close to suppressing even elementary rights of dissent. It is all too often only through struggle against the dead weight of state and territorialized power that a different right to the city can be asserted. In many cities the homeless find that struggle to be at the very core of their everyday lives. To them the injustice is palpable, while to the rest of society they are simply categorized as a public nuisance and administered their just desserts accordingly (Mitchell 2003). While the rights in this case may theoretically be equal, the force exercised to determine outcomes is invariably lop-sided.

The capitalistic logic of power, on the other hand, rests upon a quite different conception of rights, based in private property and individual ownership. To live in a capitalist society is to accept or submit to that bundle of rights necessary for capital accumulation and market exchange to proceed in a legally justifiable and enforceable way. These were the rights that were codified in universal language in the UN Declaration of 1948. The state is supposed to act as guarantor of these rights. Though it sometimes signally fails to support them, the state needs money to maintain and enhance its power and is therefore very much, in contemporary times, at the mercy of the capitalistic logic of power (even when state influence is not bought outright, in many instances quite legally, through the electoral corruptions of money power). But, reciprocally, capital needs the state to protect the rights to private property and the profit rate.

We live, therefore, in a society in which the inalienable rights of individuals to private property and the profit rate trump any other conception of inalienable rights you can think of. Defenders of this régime of rights plausibly argue that it encourages "*bourgeois* virtues," without which everyone in the world would be far worse off. These include personal and individual responsibility, independence from state interference (which often places this régime of rights in severe opposition to those defined within the state), equality of opportunity in the market and before the law, rewards for initiative and entrepreneurial endeavors, care for oneself and one's own, and an open market place that allows for wide-ranging freedoms of choice of both contract and exchange. This system of rights appears even more persuasive when extended to the right of private property in one's own body (which underpins the right of the individual to freely contract to sell his or her labor power as well as to be treated with dignity and respect and to be free from bodily coercions such as slavery) and the right to freedom of thought, of expression, and of speech. These derivative rights are appealing. Many of us rely heavily upon them. But we do so much as beggars live off the crumbs from the rich man's table.

My objection to this régime of rights is quite simple: to accept it is to accept that we have no alternative except to live under a régime of endless capital accumulation and economic growth no matter what the social, ecological, or political consequences. It also implies that this endless capital accumulation must be geographically expanded by extension of such rights across the globe. This is exactly what neoliberal globalization and its institutional framework, such as that of the World Trade Organization, the IMF, and the World Bank has accomplished. Under such a régime, imperialism of some sort is unavoidable (Harvey 2003), and the inalienable rights of private property and the profit rate will be universally established. These are the rights that are depicted as standing for goodness in a sea of evil.

It's no wonder those of wealth and power support such rights and freedoms while seeking to persuade us of their universality and their goodness. Thirty years of neoliberal freedoms have brought us immense concentrations of corporate power in energy, the media, pharmaceuticals, transportation, and even retailing (look at Walmart). The freedom of the market that George W. Bush proclaims as the high point of human aspiration towards individual liberty turns out to be nothing more than the convenient means to spread corporate monopoly power and Coca-Cola everywhere without constraint.[1] In the United States, to take a paradigmatic case, it has also permitted the top 1 per cent of income earners to raise their proportionate claim on the national income from less than 8 per cent in the 1970s to close to 20 per cent today. Even more dramatically the top 0.1 per cent of income earners increased their share from 2 to over 6 per cent of the national income between 1978 and 1998 (today it will surely be even greater) (Dumenil and Levy 2004: 4; Task force on Inequality and American Democracy 2004: 3). Figures on the distribution of wealth look even worse. Neoliberalism has simply been about the restoration of class power to a small élite of CEOs and financiers. With disproportionate influence over the media and the political process this élite seeks to persuade us as to how better off we all are under a neoliberal régime of political-economic power. And for them, living comfortably in their gilded ghettos, the world is indeed a better place. Thrasymachus, it seems, was right all along; or, as Polanyi might have put it, rights and freedoms have been conferred on those "whose income, leisure and security need no enhancing, [leaving] a mere pittance of liberty for the people" (Polanyi 1957: 257).

The remarkable increase in talk of freedom and of rights during the short history of neoliberalism in effect has displaced ideals of democracy as central issues in political struggle (Bartholomew and Breakspear 2003). Oppositional politics to global capitalism are increasingly channeled, particularly through the rapidly proliferating influence of non-governmental organizations, into paths in which individual rights and freedoms are seen as fundamental while other forms of social solidarity become less salient. This discursive, ideological, and political shift fits naturally with a neoliberal belief that, as Margaret Thatcher famously put it, "there is no such thing as society, only individuals and their families." It also reflects a changing structure of political power in which the executive and judicial branches of government gain at the expense of the institutions of representative social democracy.

But I think it unwise to abandon all talk of freedom and of rights to the neoliberal hegemony. There is a battle to be fought not only over what rights and freedoms shall be invoked in particular situations but how universal conceptions of rights and of freedoms shall be constructed. There is a huge global movement for global justice that clearly sees the nature of the problem even as it struggles to identify viable alternatives (Gills 2000; Bello 2002;

Cavanagh 2002). The bundle of rights and freedoms now available to us, and the social processes in which they are embedded, need to be challenged at all levels. They produce cities marked and marred by inequality, alienation, and injustice. In response, urban social movements arise that oppose or support the endless accumulation of capital and the conception of rights and freedom embedded therein. If a different right to the city is to be asserted and a different version of the urban process is to be constructed, then it is through the fires of specific urban struggles that any new conception will be forged.

The city, the noted urban sociologist Robert Park once wrote, is:

> man's most consistent and on the whole, his most successful attempt to remake the world he lives in more after his heart's desire. But, if the city is the world which man created, it is the world in which he is henceforth condemned to live. Thus, indirectly, and without any clear sense of the nature of his task, in making the city man has remade himself.
>
> (Park and Turner 1967: 3)

The city can be judged and understood only in relation to what I, you, we, and—lest we forget—"they" desire. If the city does not accord with those desires then it must be changed. The right to the city "cannot be conceived of as a simple visiting right or as a return to traditional cities." On the contrary, "it can only be formulated as a transformed and renewed right to urban life" (Lefebvre *et al.* 1996: 158). The freedom of the city is, therefore, far more than a right of access to what already exists: it is a right to change it more after our hearts' desire. But if Park is right—that in remaking the city we remake ourselves—then we must evaluate continuously what we might be making of ourselves as well as of others as the urban process evolves. If we find our lives to be too stressful, alienating, or just too plain uncomfortable and unrewarding, then we have the right to change course and seek to remake ourselves in another image by constructing a qualitatively different kind of city. The question of what kind of city we desire becomes inseparable from what kind of people we want to become. The freedom to make and remake ourselves and our cities in this way is, I hold, one of the most precious of all human rights.

So what, then, do I and others do if we determine that the city does not conform to our hearts' desire, that, for example, we are not remaking ourselves in sustainable, emancipatory or even "civilized" ways? How, in short, can the right to the city be exercised by changing urban life? Henri Lefebvre's answer is simple in its essence: through social mobilization and collective political/social struggle (Lefebvre 2003). But what vision do I or social movements construct to guide us in our struggles? And how can those political struggles be waged in such a way as to ensure positive outcomes rather than a descent into endless violence? One thing is clear: we cannot let fear of the latter lead us into cowering and mindless passivity. Conflict avoidance is no answer. To slide back into that is to disengage with what urbanization is all about and thereby lose any prospect of exercising any right to the city whatsoever.

This is precisely the point at which Fainstein's conception of the Just City falters. From the start, it delimits its scope to acting within the existing capitalist régime of rights and freedoms and is thus constrained to mitigating the worst outcomes at the margins of an unjust system (Fainstein 1999; Campbell 2003). Consonant with this deference to private property rights and market freedoms, Fainstein's emphasis on the discursive and inspirational role of the Just City avoids the necessity for outright conflict and struggle. Certainly, her proposal incorporates disagreement and debate, but these differences are ultimately to be harmoniously resolved, so to speak, over a cup of cappuccino at a sidewalk café. What we really need to do is understand something about the nature of conflict, which emerges from a consideration of utopian conceptions.

Most of the projects and plans we designate as "utopian" are fixed and formal designs. They are what I call "utopias of spatial form"—the planned cities and communities that have through the ages beguiled us into thinking that history may stop, that harmony will be established, that human desires will for once and for all be fully satiated if not happily realized (Harvey 2000). But history and change cannot be erased by superimposing a spatial form that locks down all desire for novelty and difference. All such utopias of spatial form end up being repressive of human desire. And to the degree that they have been implemented, the results have been far more authoritarian and repressive than emancipatory. Why, then, do we still hanker after such utopias and in what ways can this tradition be mobilized to a more open purpose? Louis Marin provides an interesting gloss here. Utopias of spatial form amount, he suggests, to a form of "spatial play" (Marin 1984). From this perspective we see the immense variety of spatial forms incorporated into different utopian plans as experimental suggestions as to how we might re-shape urban spaces more to our hearts' desire or, more cogently, to realize a certain social aim, such as greater gender equality, ecological sustainability, cultural diversity, or whatever. Conversely, we also learn to see many of the existing spaces of the city as potential sites—dubbed "heterotopic" by Lefebvre as well as by Foucault (Foucault 1986; Lefebvre 2003)—that can provide socio-spatial bases within which experiments into different modes of urban living can arise and from which struggles can be waged to build a different kind of city.

There is, however, a marked contrast between this spatial specification of urban alternatives and what I call utopias of social process (Harvey 2000). In this latter case, we presume that some social process will lead to the "promised land." In recent times, for example, neoliberal theorists (building on the liberal tradition that goes back to John Locke and Adam Smith) have sought to persuade us that freedoms of the market will bring us all wealth, security, and happiness and that we will all end up happily dwelling in the cities of our dreams. Against that are set a whole range of radical and revolutionary thinkers who have claimed that social or class struggle will eventually lead us to the perfections of communism, socialism, anarchism, feminism, ecologism, or whatever. Such utopian schemes of social process seem just as fatally flawed, both in theory as in practice. In part, the problem is that such schemas abstract entirely from the problems that arise when spatial structures get created on the ground. The territoriality of political power and organization is viewed as neutral in human affairs (when we know in practice that spatial forms are constitutive of social relations). Such frameworks of thought ignore what happens when walls, bridges, and doors become frameworks for social action and bases for discriminations. If utopias of spatial form are found wanting because they seek to suppress the force of historical change, then utopias of social process are equally at fault because they deny the constitutive significance of spatial organization materialized on the ground. Failure to grapple seriously with the problematics of urban processes leads to fatal mistakes. Socialists, Lefebvre rightly complained, found themselves "armed with nothing but childish concepts and ideologies" (Lefebvre 2003: 110) when it came to understanding urban processes. Why can we not devise a utopianism of spatio-temporal process, a dialectical utopianism that combines the idea of radical changes in both space and time to fashion an entirely different imagination of what city life could be about (Harvey 2000; Lefebvre 2003)?

This notion of dialectical utopianism seems to me to be much superior to the idea that somehow or other we are going to arrive at some harmonious endpoint. I was once invited by the Episcopalian bishop of Philadelphia to speak at his annual conference. I decided to talk about this notion of dialectical utopianism to the audience comprised primarily of theologians. They were enthusiastic about the concept, stating that the Christian concept of paradise is perpetually challenged by the fact that it is so boring that nobody wants to go there. There ensued much discussion about reframing Christian paradise in this dialectical manner to make it more appealing. A static endpoint is not desirable. Alfred North Whitehead captured this point well in his phrase: "nature is about perpetual exploration of novelty" (Whitehead *et al.* 1978: 33). As a part of nature, it seems to me that human beings are very much about the perpetual exploration of novelty. The capitalist class has understood this very well over the last 30 years, and they have used the perpetual search for novelty, new novelties, and novel novelties all the way down the line to accumulate capital and reconstitute themselves as a class. In the meantime, the Left has been focused on harmony. It proposed that we all go to Christian paradise, which is so boring nobody wants to go there. In developing visions of the city of our hearts' desire, it is essential to address this question of novelty and how novelty can be built in to our visions. But since novelty always arises out of conflict, a Just City has to be about fierce conflict all of the time.

The struggle for a Just City is a crucial battleground in the redefinition of rights. Rights and freedoms are, of course, rarely, if ever, willingly surrendered.

Remember, as Marx wrote, "between equal rights, force decides." This does not necessarily mean violence (though, sadly, it often comes down to that). But it does mean the mobilization of sufficient power through political organization as well as in the streets to change things.

No social order, said Saint-Simon, can change without the lineaments of the new already being latently present within the existing state of things. Revolutions are not total breaks, but they do turn things upside down. Derivative rights (like the right to be treated with dignity) should become fundamental, and fundamental rights (of private property and the profit rate) should become derivative. Was this not the traditional aim of democratic socialism as it sought, with some success, to use the territoriality of political power to regulate and tame the rights of capital? There are, it also turns out, contradictions within the capitalist package of rights that can be exploited for political gain. What would have happened to global capitalism and urban life had the UN Declaration's clauses on the derivative rights of labor (to a secure job, reasonable living standards, and the right to organize) been rigorously enforced? But old rights can be resurrected and new rights can be defined: like the right to the city which, as stated above, is not merely a right of access to what the property speculators and state planners define, but an active right to make the city more in accord with our hearts' desire, and to re-make ourselves thereby in a different image.

The right to the city cannot be construed simply as an individualized right. It demands a collective effort and the shaping of a collective politics around social solidarities. Neoliberalism has, however, changed the rules of the political game. Governance has displaced government; rights and freedoms are prioritized over democracy; law and public–private partnerships lacking in transparency have displaced representational democratic institutions; the anarchy of the market and competitive entrepreneurialism replace deliberative capacities based in social solidarities. Oppositional cultures have had to adapt to these new rules and find new ways to challenge the hegemony of the existing order. They have learned to insert themselves into structures of governance sometimes with powerful effects (as on many environmental issues). A vast range of innovations and experiments with collective forms of democratic governance and communal decision-making have emerged in urban settings in recent years (Montgomery 2003), including the participatory budgeting experiment in Pôrto Alegre (see Maricato this volume for more on the larger context of this experiment), the many municipalities that have taken seriously the Agenda 21 ideals of sustainable cities, the neighborhood committees and voluntary associations that increasingly take charge of public and neighborhood spaces, and the heterotopic islands of difference that exclude corporate powers (such as Wal-Mart) and construct local economic exchange systems or sustainable communities. The decentralization of power that neoliberalism demands has opened a space for all sorts of local initiatives to flourish in ways that are far more consistent with an image of decentralized socialism or of social anarchism than of tight bureaucratized centralized planning and control. The innovations are out there. The problem is how to bring them all together to construct a viable alternative to free-market neoliberalism.

The creation of a new urban commons, a public sphere of active democratic participation, requires the rolling back of that huge wave of privatization that has been the mantra of a destructive neoliberalism in the last few years. We must imagine a more inclusive even if continuously fractious city based not only upon a different ordering of rights but upon different political-economic practices. Individualized rights to be treated with dignity as a human being and to freedoms of expression are too precious to be set aside, but to these we must add the right to adequate life chances for all, to elementary material supports, to inclusion, and to difference. The task, as Polanyi (1957: 249–58) has suggested, is to expand the spheres of freedom and of rights beyond the narrow confines within which neoliberalism confines them. The right to the city is an active right to make the city different, to shape it more in accord with our collective needs and desires and so re-make our daily lives, to re-shape our architectural practices (as it were), and to define an alternative way of simply being human. If our urban world has been imagined and made, then it can be re-imagined and re-made.

But it is here that the conception of the right to the city takes on another gloss. It was in the streets that the Czechs liberated themselves from oppressive forms of governance in 1989, it was in Tiananmen Square that the Chinese student movement sought to establish an alternative definition of rights, it was through massive street demonstrations that the Vietnam War was pushed to closure, and it was in the streets that millions protested against the prospect of U.S. imperialist intervention in Iraq on February 15, 2003. It has been in the streets of Seattle, Genoa, Melbourne, Quebec City, and Bangkok where the inalienable rights of private property and of the profit rate have been challenged. "If," says Don Mitchell, "the right to the city is a cry and a demand, then it is only a cry that is heard and a demand that has force to the degree that there is a space from and within which this cry and demand is visible. In public space—on street corners or in parks, in the streets during riots and demonstrations—political organizations can represent themselves to a larger population, and through this representation give their cries and their demands some force. By claiming space in public, by creating public spaces, social groups themselves become public" (Mitchell 2003: 12).

The inalienable right to the city rests upon the capacity to force open spaces of the city to protest and contention, to create unmediated public spaces, so that the cauldron of urban life can become a catalytic site from which new conceptions and configurations of urban living can be devised and out of which new and less damaging conceptions of rights can be constructed. The right to the Just City is not a gift. It has to be seized by political movement.

NOTE

1 "Freedom is the Almighty's gift to every man and woman in this world" and "as the greatest power on earth we have an obligation to help the spread of freedom" (speech given by George W. Bush on April 13, 2004 quoted in Sanger 2004). In the context of Iraq, Paul Bremer, head of the Coalition Provisional Authority, promulgated orders in 2003 that gave substance to this notion of freedom, including "the full privatization of public enterprises, full ownership rights by foreign firms of Iraqi businesses, full repatriation of foreign profits . . . the opening of Iraq's bank to foreign control, national treatment for foreign companies and . . . the elimination of nearly all trade barriers" (Juhasz, 2004).

REFERENCES

Bartholomew, A. and Breakspear, J. (2003) "Human Rights as Swords of Empire," in *Socialist Register*, 124–45, London: Merlin Press.

Bello, Walden F. (2002) *Deglobalization: Ideas for a New World Economy*, London: Zed Books.

Campbell, S. (2003) In Scott Campbell and Susan S. Fainstein (eds) *Readings in Planning Theory* (2nd edition), Malden: Blackwell, 435–458.

Cavanagh, J. and International Forum on Globalization Alternatives Task Force (2002) *Alternatives to Economic Globalization: A Better World Is Possible*, San Francisco: Berrett–Koehler.

Dumenil, G. and Levy, D. (2004) "Neo-Liberal Dynamics: A New Phase?" in van der Pijl, K., Assasi, L., and Wigan, D. (eds) *Global Regulation: Managing Crises after the Imperial Turn*, Houndsmill, Basingstoke: Palgrave.

Fainstein, Susan S. (1999) "Can We Make the Cities We Want?" in Beauregard, R.A., and Body–Gendrot, S. (eds) *The Urban Moment: Cosmopolitan Essays on the Late-20th-Century City*, Thousand Oaks, Calif.: Sage Publications.

Fainstein, Susan S. and Campbell, S. (eds) (2003) *Readings in Planning Theory*, Oxford: Blackwell Publishers.

Foucault, M. (1986) "Of Other Spaces," *Diacritics* 16(1), pp. 22–7.

Foucault, M. and Gordon, C. (1980) *Power/Knowledge: Selected Interviews and Other Writings, 1972–1977*, Brighton, Sussex England: Harvester.

Gills, B.K. (2000) *Globalization and the Politics of Resistance*, New York: St. Martin's Press.

Harvey, D. (1996) *Justice, Nature, and the Geography of Difference*, Cambridge, Mass.: Blackwell Publishers.

—— (2000) *Spaces of Hope*, Berkeley: University of California Press.

—— (2003) *The New Imperialism*, Oxford; New York: Oxford University Press.

—— (2006) "Spaces of Global Capitalism: Towards a Theory of Uneven Geographical Development," New York: Verso.

Juhasz, A. (2004) "Ambitions of Empire: The Bush Administration Economic Plan for Iraq (and Beyond)," *Left Turn Magazine*, 12, February/March.

Lefebvre, H. (2003) *The Urban Revolution*, Minneapolis: University of Minnesota Press.

Lefebvre, H., Kofman, E., and Lebas, E. (1996) *Writings on Cities*, Cambridge, Mass.: Blackwell.

Marin, L. (1984) *Utopics: Spatial Play*, Atlantic Highlands, NJ: Humanities Press.

Mitchell, D. (2003) *The Right to the City: Social Justice and the Fight for Public Space*, New York: Guilford Press.

Montgomery, M., National Research Council (U.S.) Committee on Population, and National Research Council (U.S.) Division of Behavioral and Social Sciences and Education (2003) *Cities Transformed: Demographic Change and Its Implications in the Developing World*, Washington, D.C.: National Academy Press.

Park, R.E. and Turner, R.H. (1967) *On Social Control and Collective Behavior, Selected Papers*, Chicago: University of Chicago Press.

Plato (1955) *The Republic*, Baltimore, MD.: Penguin Books.

Polanyi, K. (1957) *The Great Transformation*, Boston: Beacon Press.

Rawls, J. (1971) *A Theory of Justice*, Cambridge, Mass.: Belknap Press of Harvard University Press.

Sanger, D. (2004) "Making a Case For a Mission," *New York Times*, A1, April 14.

Task Force on Inequality and American Democracy (2004) *American Democracy in an Age of Rising Inequality*, American Political Science Association.

Walzer, M. (1983) *Spheres of Justice: A Defense of Pluralism and Equality*, New York: Basic Books.

Whitehead, A.N., Griffin, D.R., and Sherburne, D.W. (1978) *Process and Reality: An Essay in Cosmology*, New York: Free Press.

3 Discursive planning
Social justice as discourse

Frank Fischer

Over the past decade there has been something of an acrimonious debate in urban planning theory between those who have advanced a communicative action or argumentative approach and those who have sought to reaffirm a more traditional political-economic orientation which has, among other things, emphasized social justice. In large part, those who support the traditional political-economic approach appear to worry that they have lost ground to a newer democratic-deliberative orientation based on the work of planners influenced by the epistemological contributions of Habermas. Some have even expressed concern that the communicative approach is *becoming*, if it has not already *become*, the dominant paradigm in planning theory. For them, the communications approach leaves out other more traditional questions, in particular concerns about social justice. The purpose of this chapter is to enter this debate—at times unfriendly and unproductive—to show that these two orientations are often closer to one another than first appears to be the case. Indeed, both social justice and a discursively oriented communicative approach, it is argued, are at important points necessarily dependent on one another. Seeking to circumvent this theoretical standoff, the discussion illustrates how the work of the discursive planner and the political-economist can be brought together in a more encompassing normative framework based on practical discourse. Toward this end, we begin with a brief outline of the communicative action approach, then turn to what their critics have to say about it.

COMMUNICATIVE ACTION IN PLANNING THEORY: LANGUAGE, ARGUMENTATION, AND DISCURSIVE PRACTICES

The communication model or "argumentative turn" has over the past 15 years become an important theoretical orientation in the fields of urban planning and policy analysis, although scarcely the dominant orientation (Fischer and Forester 1993; Sager 1994).[1] Led in planning by theorists such as Forester, Healey, Innes, Hoch, and Throgmorton, the objective of the approach is to examine the way the planner communicatively engages in political and professional practices. Focusing on the relationships between knowledge and power, as Ploger (2001: 221) puts it, these theorists examine the "acts of power such as words in use, argumentation in action, as well as gestures, emotions, passions, and morals representing institutional politics and ways of thinking." At the same time, it has also involved the analysis of forms of participatory or collaborative planning (Healey 1997).

These scholars take one of the primary activities of the planner to be the facilitation of the processes of deliberation. Towards this end, they propose an argumentative or discursive analysis of planning practices (Forester 1999). From this perspective, an important source of knowledge about planning is the discursive activities of planners at work: their written and spoken words in their deliberations, plans, and other relevant documents. In Forester's view, no aspect of public planning is more important than communication. Focusing on public disputes and contested information in the face of power, his focus is on "words in practice." Recognizing that the institutions in which planners work are a basic source of distorted communication, critical analysis must focus on "the politics (status, strategies, effects, and implications) of who says what, when and how in planning-related organizations" (Forester 1999: 53; 1988).

For communicative action theorists, the investigation of such discursive practices emphasizes the individual professional in local settings, whether the town meeting or the office of the department of planning. They want to know what actually happens when planners speak and listen in the course of their practical activities (Forester 1999: 49), in particular how these communicative interactions are interpreted through basic systems of meaning (Healey 1997: 49). The goal is to help planners learn to critically reflect on their own discursive practices, in particular their ways of arguing. They seek to discover the messages that are latently conveyed through the language they use, as these are often the hidden meanings in what otherwise appear to be clear statements of intent. As Ploger (2001: 221) explains it, the objective is to make policy planners and analysts "more conscious of the hidden forms of communicative power they practise (often unconsciously) in order to develop a democratic, yet rational, public communication."

Underlying this point of view, especially for Forester and Healey, is a Habermasian perspective on "distorted communication." Through descriptions of concrete cases, the goal is to make the practicing planner aware of the often manipulated or distorted nature of communications in the planning process. Such work reflects an interest "in the relationship between knowledge and power, in the potential for oppression inherent in instrumental rationality, and including a more emancipatory way of knowing" (Sandercock 1997: 96). Only when planners themselves become aware and self-critical of these often subtle power mechanisms, in this view, is it possible to develop a communicative solidarity. Toward this end, the planner's task is to work to establish more equal communicative forces of power through discursive counteractions. The planner should see him or herself not only as an agent of political-institutional power, but also as part of the effort to reform society. Forester and others have therefore labored to interweave advocate planning, social learning, and radical planning as a mode of self-critical communicative practice.

In the process, such planners seek to create discursive spaces and opportunities for more consensual modes of planning and policymaking. As Sandercock (1997: 96) puts it, the task is to assure "representations of all major points of view, equalizing information among group members, and creating conditions within group processes so that the force of argument can be the deciding factor rather than an individual's power or status in some pre-existing hierarchy." In political terms, the goal is to understand how public deliberation among the social actors affected by the consequences of policies and plans facilitates less technocratic, more democratic planning processes. Epistemologically, following Dewey as well as Habermas, the challenge is to understand how discursive processes are basic to reason itself. Such planners conceive of reason and rationality as residing in discursively negotiated and communicated understandings and intentions within communities bound together by traditions, conventions, and agreements forged through democratic deliberation (Hoch 1994: 31, 1997, 2007).

For those who adhere closely to Habermas' theory of communicative action, discourse and deliberation in the planning process can be analyzed in terms of the normative requirements of ideal speech. From this perspective, a critical approach to planning is grounded in a "universal pragmatic" that is understood to be inherent to speech. If deliberative dialogue is based on the sincerity, comprehensibility, truthfulness, and normative legitimacy of arguments, then the universal principles of ideal speech can be employed to evaluate speech claims as they unfold through inter-subjective communication (Hoch 1997; 2007). These universal criteria can serve as "practical moral guidelines to a democratic, intersubjective communication and as premises to uncoerced reason as the outcome of discussion" (Ploger 2001: 221). It is not, as is too often assumed, that these theorists believe that the planning process can be redesigned in terms of the requirements of ideal speech, but rather that its criteria can be used as counter-factual principles that can serve as criteria for evaluating the deliberations that do take place.

THE COMMUNICATIVE TURN: DOMINANT PARADIGM?

The debate in question appears to have started with an essay published in the planning journal by Judith Innes (1995). In that article, Innes suggested that the communicative approach to planning had the potential to become the dominant orientation. Although she did not state that it *is* the case, she did say that it might become the dominant paradigm. Given the misunderstandings to which it gave rise, the statement was unfortunate, which Innes has herself conceded. Appearing in a leading planning journal, it nonetheless set off a significant outcry among those representing more traditional approaches, in particular the critical political-economically oriented planners.

In important instances, the response offered has been a call for an emphasis on social justice.

Starting from the assumption that the communicatively-oriented approach harbors—explicitly or implicitly—the desire to become hegemonic, the critics launched several interrelated salvos against the theory and its proponents (Huxley and Yiftachel 2000; Yiftachel and Huxley 2000). Some have argued that the approach is not encompassing enough to cover all of the aspects of the field of planning. As Mazza (1995) has argued, the professional planner is typically engaged in a wide range of activities not well accounted for by the communications approach. For one thing, it would seem to neglect the empirical activities of the planner as researcher—the measurement of settlement patterns, transportation distances, or the distribution of urban income, and the like. In point of fact, there is no question about this, although "neglect" might not be the appropriate word. In reply, communicative theorists argue that they are not against such work; they are rather focusing on a different dimension of planning, one they see as neglected. They have also added that while they are not focused on the kinds of traditional empirical activities that Mazza has in mind, communicative theorists are very much interested in how planners handle empirical information about such matters in the policy planning process.

Closely related, others have argued that the proponents of communication and deliberation have ignored the traditional topic of the field, namely cities and urban areas (Fainstein 2000). Again, this is true, although it tends to overlook the fact that communicative theorists have typically focused on planners working in the city planning department and surely much of the talk there is about the city and its residents. It might not be of interest to everyone, but the stress has been on understanding the communicative interactions of planners in this context, as well as how planners relate to the relevant urban groups, citizens, administrators, and politicians. Concomitantly, this has led various scholars working along these lines to focus on ways to facilitate public deliberations among such groups.

Moreover, other critics argue that an emphasis on the communicative activities of the working planner fails to constitute a critical approach to theory and practice. Given that most practitioners work for city agencies of one type or another, the conversations recorded in a government agency are by their nature a form of communication influenced—if not distorted—by administrative and political interests embedded in the organizational structures and processes. One can learn how successful planners strategically use language to persuade, but this scarcely constitutes an ideal communicative setting capable of generating an authentic consensus (Throgmorton 1996, 2003). Toward this end, the critics ask how planners working in the "belly of the beast" can actually employ such principles. Are not their activities constrained by the ideas and discursive practices embedded in the institutions for which they work? Does this not indicate a lack of understanding and neglect of power? Indeed, it would not even be wrong to argue that many of their professional practices are a product of the needs and interests of planning agencies. The discipline itself emerged to serve this function. In this light, no one can reasonably dispute the fact that most conventional planners and policy analysts, as public servants working for the state apparatus, mainly reproduce the existing power relations embedded in the institutions for which they work (Yiftachel 1998). For this reason, some have asked if planners can change the rationality of policy and planning practices by introducing the language and disciplinary discourses of "the noble planner" (Flyvbjerg 1996: 388).

Indeed, various critics portray communicative action theorists as naïve in believing that planners can change much of anything, being the products of larger political-economic forces. In statements that often curiously seem to denigrate planning as a profession, some have argued that the focus on planning activities is at best a focus on trivial activities, at worst a misunderstanding of urban processes. It simply fails to understand how change comes about. Further, communicative action theory, it is said, understands the planner as an intermediary negotiator and facilitator of communication. They seem to think, it is argued, that as soon as communicative inequalities are eliminated the structural differences will disappear. They write as if the goal is to make ideal speech the objective of planning. Communication, they argue, is not of much interest since language and rhetoric cannot affect anything significant on the ground. Being empty, it offers no critique.

One can agree with important aspects of the critique. Many of the critics' points are valid, although they go too far in their assertions. The urgency of the points depends on the possibility of the communicative action approach to become the dominant paradigm. In reality, there is little worry of this. Focused only on a specific set of activities involved in planning, it is far too narrow to play that role. Second, nobody involved with the approach actually thinks that this is what it is about. Although one can read her words differently, Innes has retracted the statement and others such as Forester and Hoch have formally disassociated themselves from such intentions. If the approach is seen only as a particular topic among many, the communicative theorists can be credited with making a contribution that enriches the field. It is one that is helpful in teaching planners to better understand the nature of the job and how to be more effective, especially communicatively. Given the importance of the issue to the teaching of planning, the earlier neglect of the perspective can be seen as something of a curiosity.

One might be inclined to leave the issue here. The critics score some points, but there is little reason to worry about their worst fears. There is no chance of—and little interest in—elevating the communicative action approach to the dominant paradigm. But there is a deeper concern embedded in this debate that should not go unattended. An issue that relates to both positions, although in different ways, has to do with the role of language and discourse. Underlying much of the critique of communicative action is an implicit—and at times explicit—rejection of the discursive turn to language. Some, in fact, seem to still think that language is an empty vehicle that only reflects reality and thus has no independent effect on the world. Lauria and Whelan (1995), for example, have argued that the turn to language is "substantively vacuous." In their view, the focus on language and discourse is not a useful way to understand the "articulation" of issues in planning. Feldman (1997), for another, simply writes off the emphasis on talk as some sort of "comedy."

PLANNING AS DISCIPLINARY DISCOURSE: THE DISCURSIVE APPROACH

A deliberatively oriented communications approach to planning and policy involves much more than just talking and listening. Grounded in a discursive understanding of knowledge, it is anchored to a much deeper focus on discourse and deliberation. Emerging in large part as a response to the dominant technical/technocratic approach to planning and policy analysis, the discursive orientation rests on a recognition of the critical role of language and discourse in human affairs. From a poststructural perspective, it examines the ways in which language and discursive practices actually shape and control the social world. Indeed, following Foucault, it demonstrates how disciplinary expertise and its discourses became the quintessential form of the modern system of power and social regulation. Planning, in its various configurations, is in fact one form of such discursive power.[2]

Discourse is basic to the production and reproduction of society. Hajer (1995) has defined it as "a specific ensemble of ideas, concepts, and categorizations that are produced, reproduced and transformed to give meaning to physical and social relations." Discourse and discursive practices, in this way, circumscribe the range of subjects and objects through which people experience the world, specify the views that can be legitimately accepted as knowledge, and constitute the actors taken to be the agents of knowledge. As Shapiro (1981: 231) has put it, social actors—planners, politicians, and citizens—do not "invent language or meanings in [their] typical speech" and, as such, "end up buying into a model of political relations in almost everything [they] say without making a prior, deliberate evaluation of the purchasing decision." Thus, given that the languages of politics inscribe the meanings of public problems, a policy or plan to deal with them is not only expressed in words, it is literally constructed through the language(s) in which it is described.

To offer a simple but clear example, it makes a significant difference whether policy deliberations over drug addiction are framed in a medical or legal discourse. That is, to say something one way rather than another is to also explicitly say a whole host of other things, which will be grasped by some and not by others. In this way, approaches to public intervention are constructed from the history, traditions, attitudes, and beliefs of a people encapsulated and codified in the terms of its discourses. A discursive approach not only examines these discourses, but seeks to determine which political forces lead to their construction.

Or to take an instance that relates more specifically to the present discussion, nobody has ever seen a political economy, despite the fact that we talk about them as if we had. A political-economic system is a linguistic concept discursively invented and employed to describe a set of relationships that we can only partly experience—one goes to the bank, speaks with the director of a factory, reads the financial pages, or goes to the marketplace. But one never sees an entire political-economic system. While we can encounter parts of it or discover its effects, the system itself remains a set of formal and informal relationships that can only be constructed and discussed through language and argumentation.

From this perspective, the institutions of a political economy are embedded in specific discursive practices that produce and reproduce ideas and behaviors basic to the understanding and functioning of the system, including understandings of social justice and fair treatment. Over time, interpretations of fairness become unreflectively taken for granted; they are scarcely noted by the actors who employ them. As generally accepted presuppositions, they become embedded in the institutional deliberations and practices that produce and govern basic political-economic relations, including who is entitled to what. Through such discursive delimitations of a political economy—its interrelated structures and functions—the norms and standards for establishing conceptualizations used to understand and interpret the actions of the system are established, including its consequences for social equality and the distribution of income. Language thus "becomes part of data analysis for inquiry, rather than just a tool for speaking about an extra-linguistic reality" (Shapiro 1981).

While a society is typically constructed around a dominant discourse—say that of capitalism, including its understandings of justice and democracy—there are always other discourses in play. There are oppositional discourses, such as socialism or radical environmentalism, but these operate on the margins of the social order. Within the dominant system itself, there are subdiscourses related to the different sectors of society that need to operate in relationship to one another—political, economic, religious discourses, among others. To regulate the interactions of these subsystems, various disciplinary discourses emerge to coordinate and guide them. Economics, for example, develops a body of knowledge grounded in the principles and ideology of capitalism, as do disciplines such as law and management. Within this broad discursive framework, politics is advanced through a process of argumentation, drawing on and combining various threads of the available discourse. And it is here that the discursive approach enters the stage.

Recognizing the discursive nature of societal interaction, the discursive approach challenges the technical-rationalist orientation's treatment of plans and policies as apolitically value-neutral (Hajer and Wagenaar 2003). Instead, they are recognized to always be grounded in particular ideological assumptions. Without neglecting techno-empirical analysis *per se*, the discursive approach seeks to understand such analysis in terms of the normative assumptions it rest on and/or supports. It is thus grounded in the argumentative struggles that define a society at any particular time. Based on a postempiricist epistemology, it recognizes that social and political life is embedded in a web of social meanings produced and reproduced through discursive practices.

Rather than "reality" itself, postempiricism focuses on science's *account* of reality (Fischer 2003). This is not to assert that there are no real and separate objects of inquiry independent of the investigators. Instead of the objects and their properties *per se*, the focus is on the vocabularies and concepts used by social actors to know and represent them. The goal is to understand how these varying cognitive elements interact discursively to shape that which comes to be taken as knowledge. Given the perspectival nature of the categories through which social and political phenomena are observed, knowledge of a social object or process can be seen as something that emerges more from discursive—dialectic—interactions among competing interpretations. This opens the door to interpretive methods, rhetorical argumentation, and the analysis of deliberative processes (Throgmorton 1996, 2003).

Moving beyond epistemology, in the practical world of action the discursive approach turns to the medium of argumentation. Rather than concentrating on empirical analysis, the discursive planner or policy analyst examines the ways that a mix of data, technical tools, concepts, and theories structure practical arguments (Majone 1989). The approach thus seeks to bring a wider range of contextually relative evidence to bear on the arguments under investigation. Toward this end, it draws on the normative logic of practical discourse.[3] Discursive deliberation becomes the essential method for considering and sorting out the components of an argument. This is the case in rigorous disciplinary analysis: the experts assemble evidence and deliberate its meaning from different methodological perspectives, including its action oriented implications. By cognitive extension, albeit in less rigorous form, citizens are recognized to do the same with their own "ordinary knowledge" (Lindblom and Cohen 1979; Lindblom 1990).[4] A primary task of the discursive planner is to develop methods of deliberation relevant for helping citizens to do this (Fischer 2003, 2009).

The discursive approach is thus a response to an epistemological gap. Whereas some critics speak as if the communicative orientation hinders or diverts attention away from the task of generating empirical knowledge for planning and problem-solving, discursive planning emerges to confront the fact that the professions have yet to develop such knowledge. There is no such body of knowledge, and not for a lack of empirical analysis of pressing problems. Rather, the problem has to do with the nature of the social world and our ability to understand the complex phenomena that define it. While this is not to say there is no expert knowledge about social and economic problems, it is to recognize that what we do have is only partial and can be applied only through a discursive examination of the uncertainties of this knowledge in terms of a range of other empirical and normative considerations basic to the context to which it is to be applied (Flyvbjerg 2000). The discursive planner thus seeks a way out of a dilemma. It is an action-oriented attempt to find, understand, and interpret the partial knowledge that we have in the practical context to which is to be applied. This might not be planning theory *per se*, but it has to be a very important part of a planning discipline that hopes to be able to usefully apply its knowledge.

SOCIAL JUSTICE AS DISCOURSE

How, then, does this perspective apply to social justice and an attempt to identify the Just City? In an effort to answer this question, it is important to first look at the specific characteristics of social justice, or perhaps more appropriately the concept's lack of characteristics. Basically we discover here a concept without a clear content. At the most general level, social justice refers to the degree to which a particular society affords groups and individuals fair treatment through an impartial division of advantages and disadvantages. Beyond this, however, there is little or no agreement on the definition of what impartial share and fair treatment mean. Most people wish to inhabit a just society, generally understood as a social order that supplies some minimum level of those things that can be had, for example employment and health care. But this does not carry us very far in the effort to specify the concept's content.

Further, we find a host of competing political-economic ideologies offering various conceptions of what a just society should look like and how it would operate. Consider how differently social justice is used by members of the political left and right. For the left, a socially just society refers to a social order that features a high degree of economic and social egalitarianism (which can be pursued through policies of income redistribution, progressive taxation, and property redistribution). Social justice, in this regard, has always been a basic principle of socialism and is still today reflected in the social democratic concept of a welfare safety net. Political conservatives, at the same time, take the just society to be one that is based on the free market, seen to offer greater equal opportunities than state-directed public policies. From the capitalist perspective, social injustices are better corrected through individual initiatives, private philanthropy, and charity.

These competing political positions have led important theorists to point out that there is no objective standard of social justice. For many, it is a term of moral relativism, a position advanced by postmodernists. Some more simply reject it as meaningless. Yet others, following Machiavelli, see it as a contentless concept employed to justify the *status quo*. In this view, social justice is what those with power say it is. From this understanding, then, one can offer a theory of the Just City, but it cannot be more than one of numerous other contested positions and will be treated as such by those with different preferences. This is to say, it cannot be established once and for all by accepted criteria. One can still call for more social equality, but the argument rests on a political preference (Gans 1974).

Of what value, then, is this concept? From the deliberative perspective, the term is essentially an invitation to engage in a discourse—that is, to deliberate about the nature of equality and opportunity in a particular society and how its members might go about changing existing arrangements. Such discourse operates on two levels—one theoretical, the other practical. On the theoretical level, much of the contemporary discussion of social justice takes the philosophical work of Rawls (1971) as its starting point. Here the issue is about the distribution of values that people would pick if they were to decide—through personal deliberation—"behind a veil of ignorance" that permits them no information about their own social and economic standing. Via an elaborate process of theorizing Rawls concludes that individuals would choose a system of equal opportunity (which would involve a set of legal and political institutions that compensate for long-run trends of economic forces so as to prevent excessive concentrations of property and wealth, especially those that lead to political domination) (Rawls 2001: 44). Without necessarily agreeing or disagreeing with Rawls' conclusions, or going into the extensive discussion of the various facets of his position, one quickly realizes that this level of theorizing does not much help the planner in the real world. As the conditions for a societal thought experiment do not prevail in the world of action, the political philosopher's world is far removed from that of the professional planner.

Although theoretical discussions can help us think about social justice, the outcomes of such reflection are far too abstract to guide planning decisions. Planning, in this regard, operates between theory and action. It relies, as such, on basic urban values, largely derived from centuries of urban experience, and from social struggle in particular. Fainstein (2008: 20–22), for example, offers a list of these urban values, categorized under three basic value orientations—equality, diversity, and democracy. Equality for the planner, she argues, is pursued through standards of income distribution, special benefits provided to low-income citizens, low transportation fares, and the like. Diversity relates to the need for mixed populations and public spaces, typically pursued through zoning laws. Democracy, as she defines it, involves consultation about urban development; it takes broader interests of the city into consideration. Missing from the list, I would argue, is the value of efficiency, which comes into play at the other end of the theory-action continuum. Without a measure of attention to efficient action, few problems can be solved.

Planning, in the view here, can best be understood as the task of integrating technical efficiency and social equality. The question of social justice is in significant part a matter of how to balance these often conflicting values against one another. Despite efforts to establish an optimal relationship between equity and efficiency—such as the Pareto Optimal solution in economics—such concepts remain in tension and often only lend themselves to trade-offs (Okun

1975). One can get more social equity but often at the expense of technical efficiency; and conversely, more efficiency at the cost of social equity.

Who, then, has the legitimate authority to establish this mix of values? How is this to be done in a highly pluralist society populated by people from diverse ethnic, racial, cultural, and class backgrounds? A good deal of historical experience shows that this cannot be successfully accomplished from above.[5] The illustration *par excellence* is that of socialism. Whatever else one wants to say about socialism, it represents a long-standing attempt to plan a socially just society from above. Drawing on Marxist principles of political economy, a harmonious balance between economic efficiency and social equality was to be determined by scientific analysis of the laws of economic and social motion. Introduced by the socialist party and enforced by a centralized state, the project was not without its successes. The Soviet Union turned a peasant society into a twentieth-century world power in the course of a half century. But the history of the effort shows that its achievements in the name of social justice were obtained at the expense of other social values, democracy and civil rights in particular. The result was an array of drab social systems based on authoritarian leadership. At the risk of simplifying a complex history, it is reasonably fair to say that a large number of their citizens, given the chance, would have chosen to live under a different distribution of social values. One can also say that the failure of centralized planners to either successfully or legitimately settle these questions in the name of the people is one of the basic lessons from the experience.

If communicative planners spend too much time listening, as argued by their critics, one can surely note that socialist régimes listened too little. The story, told too many times to repeat, underscores the basic assumption of democratic government, namely that the few should not speak for the many. The experience has shown that the political leaders who have sought to guide these societies were often as fallible as the general citizenry. And it has also illustrated the problematic nature of the technocratic effort to anchor the project to a highly positivist conception of political-economic science. Given the underdetermined nature of the science, economic planners proved unable over time to effectively balance productive efficiency and social equality. Today, thanks to decades of postempiricist reinterpretation of science, the social and economic sciences in particular, we know that while political economy can help us think about particular problems, it cannot *solve* these problems in a world of complex uncertainties. Explanation and solutions, instead, necessarily come from a mix of theories, findings, specific commitment, and political judgments. In large part, a disciplinary community exists to deliberate about these interrelationships. Indeed, it is just here that the discursive perspective emerges to address this reality. Rather than an alternative epistemology *per se*, it seeks to explain what scientists actually do as they go about their business.

In the absence of both objective criteria and hard knowledge, discursive planners seek first to establish an acceptable (i.e., workable) set of criteria for proceeding in the realm of social action. Toward this end, they pursue social justice from the alternative direction—namely, by establishing the criteria and content of justice by submitting the issue to the people who have to live in the societies to which it is applied.[6] It might be the case, as socialists have argued, that parliamentary institutions have often served as a "talk shop" in support of *bourgeois* interests, but socialist planners themselves have been unable to produce legitimate plans supported by high levels of commitment and motivation. It is also the case that Western capitalist democracies have tended to err in the opposite direction; they have often generated high levels of economic and social inequality and failed to make good on commitments to minimize its effects. From the deliberative perspective, the alternative—arguably the only alternative in the absence of major societal transformation—is to make the talk shop more democratic.

The turn to democratic deliberation requires no fundamental justification; it is already one of the basic societal values. In a highly inequitable society such as the United States, however, democracy has often been narrowly defined to limit the range of possible options and outcomes. Typically, it is reduced to a form of "consultation," the administrator's constricted understanding of deliberation. The deliberative planner essentially seeks to restructure this need for consultation. Instead of a top-down process based on distorted—if not manipulated—communication, the deliberative planner seeks to introduce a more authentic process of democratic consultation. As a procedural concern, deliberation is not about social justice *per se*, but it is about how we go about deciding what people will take it to mean. Insofar as the term is defined by particular groups, especially those with power, the challenge is to agitate for more equality by extending deliberation processes to a wider range of groups, especially to those who are disadvantaged. From the perspectives of discursive politics and discourse ethics, the task is to examine who participates and under what conditions. Beyond the political élites, the effort is to include the larger community.

In the process, discursive planning recognizes that different societies and communities value things differently and that people within these communities also have different preferences. There can thus be no one conception of social justice that fits all circumstances. The approach recognizes that only a conception of social justice that has been put up for deliberation among those who have to agree to live by it can succeed. This would not only involve the broad membership of a society, but also those in its subsystems who confront particular circumstances.

The discursive orientation is not, as such, a narrow or naïve emphasis on ideal speech, as often portrayed. It is also more than just a process of listening to citizens. Instead, deliberation and the argumentative process are themselves seen as part of the political struggle. Insofar as social justice enters politics through the processes of argumentation, it is advanced as a way to help people to construct and put forth better arguments. Without confusing argumentation for politics, it seeks to facilitate the discursive struggle basic to the political process. Toward this end, it can use Habermasian principles of ideal speech to both critique particular arguments and guide the construction of such discursive arrangements, but it does so without confusing these criteria with real world situations. Indeed, the basic goal is to set out theoretically and practically the intellectual and material conditions that can make deliberation meaningful. This reflects the recognition that genuine democratic deliberation depends itself on a degree of equality, both in terms of resources and rights.

Given that existing inequalities are embedded in systems of political and economic domination, the approach pays particular attention to power relationships. It recognizes, as Healey (1997: 66) has put it, that power relations "are part of us, and they exist through us, and citizens should therefore be aware of the sources of power they can seek and develop." The greater the extent to which a planner or policy analyst can better understand the dominant power structure, including the discursive nature of its ideological politics, the greater are the chances of developing effective strategies for challenging rather than merely reproducing it. If the planner or policy analyst is aware of how to challenge the dominant power system, it is then possible to struggle to change the existing forces rather than merely reproducing them. Deliberative collaboration, in Healey's view, is seen as a way of connecting community members across disparate networks which can, in the process, change their ways of seeing, knowing, and acting. Collaborative planning, as communicative power, thus has the potential to alter both power relationships and the culture of place.[7] As an approach that seeks to reveal exploitative structures, communicative structures in this case, it can help citizens become better aware of their own interests and how to pursue them (Ploger 2001: 225).[8]

Discursive planning thus means creating spaces for democratic deliberation that offer a place for citizen participation in both goal-setting and conflict management. By asking who is privileged and who is marginalized by existing forms of governance, such an approach challenges the formal institutions to be democratic and collaborative (Healey 1997: 201). The task is to both make clear the sorts of communicative power that oppress and to facilitate the kinds of knowledge that lead to deliberative empowerment (Fischer 2009).

From this alternative perspective, the relationship between social justice and democratic deliberation is not a matter of either/or. The effort to establish and advance a legitimate conception of social justice in the practical world of planning depends on its relation to democratic discursive practices. In turn, genuine deliberation requires a significant level of social equality. The chapter can, in this regard, conclude by returning to the original dispute between deliberative planners and their critics to show how it can be more constructively approached with the assistance of the kind of normative logic of practical discourse introduced in the discussion of disciplinary discourse. It helps to clarify the way in which the debate rests in significant part on a limited understanding of the nature of a normative/practical discourse rather than altogether irreconcilable approaches.

DELIBERATING SOCIAL JUSTICE: THE LEVELS OF NORMATIVE DISCOURSE

By recognizing social justice to be a normative discourse—a discourse concerned with "what ought to be"—we can examine it in terms of the structure and logic of such a discourse. Drawing on the good reasons approach to normative discourse, as I have worked out elsewhere, the complete structure of a normative/practical argument involves four interrelated levels of discourse.[9] From this perspective, the communicative/discursive planners and their critics can be seen as talking at points about the same thing, but on different levels of discourse. The conflict between those studying planning talk and those calling for greater attention to social justice can, in this view, involve different but interrelated aspects of the same discourse. Rather than representing altogether different orientations to planning, one is focused on the micro level and the other on the macro level of a discourse about social justice.

Much of the investigation by planners who have adopted a communications approach has in fact focused on low-level practical discussions, especially when listening to planning talk in the workplace. Planners employed by city governments operate under specific political and administrative mandates and must be attentive to the constraints they impose. Social justice is here officially spelled out by those with the power to define it. The larger questions of social justice rarely emerge on the grander scale of political philosophy.[10] Indeed, they are often discursively suppressed. But this does not mean that the questions and the issues they pose are not present. Planners at this level of discussion always speak about aspects of equity, but they typically do so in more immediate pragmatic terms quite removed from questions of social philosophy.

In terms of the normative logic of practical discourse these discussions take place at the level of first-order discussion. One can identify two specific kinds of basic questions that organize deliberation at this level. The first is fundamentally technical in nature: Does a plan or policy work? Is it efficient or effective? Does it lead to unexpected consequences, and the like? These are the questions to which most of the standard empirical planning techniques are applied. An empirical assessment of the technical effectiveness of a plan of action is judged against a particular standard or norm, e.g., the affordability of a housing project, or the safety standards of a power plant.

Beyond this question, practical deliberation moves into neglected normative terrain. While most planning and policy analysis focuses on questions about efficiency and effectiveness, a critical analysis must include an evaluation of the standard(s) against which a plan is measured. This involves three additional discourses, the first of which focuses on the specific situation to which the normative criterion is applied. As the second deliberation of first-order discourse, it asks if there is anything about the particular situation that would require making an exception to the technical assessment. Are there any situational circumstances, either empirical or normative, that need to be taken into consideration? For example, is a waste incinerator too close to a dense urban population? Is the proposed housing ill-suited for the kinds of people moving into the area? Does a job-retraining program address the needs of the relevant group of unemployed workers? One can accept the empirical validity of a technical assessment, but argue that the program is inappropriate under particular circumstances. With regard to a waste incinerator, for instance, the issue is well illustrated by the politics of NIMBY, as well as struggles about environmental justice.

Both of these discourses—technical and situational—capture much of what is discussed in the planning office and represent important issues in planning practice. But they still do not deal directly with the concerns of those arguing for greater emphasis on social justice. Indeed, if the analytic approach to planning is left at this level, serious questions are raised for a critical approach to planning. As soon as we move to second-order discourse, however, these more critical questions emerge as logical extensions of lower level pragmatic issues.

Second-order discourse involves justifying first-order judgments in terms of the larger society and the normative principles upon which it rests. That is, the move beyond an assessment of the technical and situational implications of a plan directly involves probing the second-order societal consequences of adopting these judgments. The first of the two second order discourses—i.e., the third-level of discourse—requires asking if the particular plan has instrumental or contributive consequences for the society as a whole. Is there anything about the social-order—its empirical processes or normative goals—that would render the technical and situational judgments problematic? Here, for example, we confront basic questions of political economy. Does a particular job-training program offer real opportunities in the existing labor market? Or does it simply supply unjustifiable tax breaks to employers who provide the training? These are questions that require a basic political-economic analysis of the structures and functions of the social system. Only with an accurate analysis of how a society's political-economic system functions can the contribution of a particular policy or plan be evaluated.

But the acceptance of a social system itself can never be taken as given. And it is here that we arrive at the fourth level of discourse, which introduces the tasks of social and political critique. At this level we again encounter two basic questions. The first asks if the social system works in ways that live up to its own norms and values. For example, does it in fact fulfill the standards of equal opportunity and fair treatment that are said to legitimate the system? And second, are there alternative social systems that more successfully fulfill these or other desired social principles and values? We reach here the theoretical/philosophical level of a discourse about social justice. For assistance with deliberation at this level, one might well turn to the works of philosophers such as Rawls for insights about social justice, or Habermas for guidance on communicative interaction and democratic deliberation.

The point is this: A normative deliberation about social justice manifests itself across the full range of practical discourse, from the very concrete to the highly abstract. It involves, in the process, a range of discourses that rely on the most relevant methodological orientations—techno-empirical analysis, participatory observation, political economic analysis, and philosophical critique. Such methods pursue different but interrelated questions logically interconnected through a full-scale, comprehensive practical discourse. As is often the case, those operating on one level of discourse fail to see the connection of their discourse to the other discursive components of a complete normative judgment. From this perspective, we can specify the task of a critical discursive theory as the identification and explication of these connections. If we start with the critique of communicative planning, the goal of a critical communicative approach should focus on how these higher levels of normative judgment are already embedded in everyday bureaucratic discussions, albeit concealed. The critical assignment is to illustrate how those discussions are in fact about social justice, although these connections are truncated and hidden from view. The essential contribution of a critical discursive analysis, following in the traditions of Habermas and Foucault, is to elucidate the ways particular aspects of higher level discourses are embedded as assumptions in the lower, more pragmatic levels of discussion. Indeed, only after clarification of these interconnections can authentic democratic deliberation take place. A discursive approach, so conceived, is scarcely irrelevant to the questions of social justice. Indeed, a meaningful discussion of the Just City depends on it.

CONCLUSION

The chapter began with an explication of the debate between planning theory focused on communicative interactions and those pursuing a more critical political-economic approach. Although the goal has been to clarify and support a discursively oriented approach to the communicative perspective, the discussion has conceded that much of what has been undertaken from the communicative perspective has failed to identify its critical potential. When it has examined talk about equity, it has too often neglected to reveal the discussion's deeper roots in a discourse about social justice. One need not specifically single out these connections as *the* topic of research, but to fail to acknowledge the existence of such connections is to invite criticism.[11]

At the same time, I have tried to show the ways in which political economists and social justice theorists claim too much. Their questions are, to be sure, important. But they are not only far removed from the tasks of the planner—as practicing planner—they are questions without easy answers. For one thing, they are not in possession of sufficient knowledge to settle the issues they raise; for another they lack the moral authority to decide them. While they can bring relevant information and knowledge to bear on decisions about equity and justice, the decisions can only be made through a process of deliberation. Indeed, as we saw, the discursive approach emerges to deal with this very problem—in the discipline as well as the public planning process. The discursive planner, in the public context, argues for democratic deliberation with a broad spectrum of voices.

Such deliberation needs to rest on an open and transparent process of communication. Toward this end, the chapter concluded by arguing that a critical discursive approach to planning involves explicating the connections between the levels of discourse and the crucial normative assumptions that are otherwise embedded in these discourses. From such a perspective, the communicative planner needs to carefully explicate the discursive connections of everyday planning talk to the higher-level theoretical questions upon which they ultimately rest. The advocates of social justice must at the same time take into account the ways their theoretical questions are already embedded in more mundane conversations about planning. A critical approach, in short, involves establishing the connections between the practical and the theoretical. This is necessary for understanding how the norms of social equity and justice operate in and through existing institutional practices, as well as how authentic democratic deliberation can be better designed and facilitated.

NOTES

1 The term "communicative action" is drawn from Habermas' theory of communicative action. It has been used in the planning debate to refer to a specific line of research, although the term relates broadly to work on argumentation, deliberation, and discourse. I shall use the term "communicative action" to outline the debate, but later emphasize the argumentatively oriented discursive strand of the communicative orientation, referring to it as a "discursive approach." The term is used to refer to a postempiricist orientation that can contribute to planning theory, including social justice in planning theory.
2 This presentation of a deliberative (communications) approach draws on contributions from both planning and policy analysis, particularly those influenced by work on postempircist/postpositivist epistemology and deliberative democratic theory. Policy analysis and planning are generally viewed as closely related disciplines, with policy analysis often taken to be one of the methodological techniques of planning.
3 There is no easy distinction between normative and practical discourse. Sometimes the terms are used rather interchangeably. In the discussion here both are considered to be normative, concerned with what "ought to be done." We use normative to primarily refer to questions about norms, standards, and ideals, whereas practical discourse refers to a larger assessment that brings empirical and normative discourses together in pursuit of a decision about what to do.
4 Lindblom defines social scientific knowledge as a refined version of "ordinary knowledge." In this respect, there is no fundamental epistemological differentiation between rigorous ordinary knowledge and the everyday ordinary knowledge of the citizen.
5 For Plato, who first sought to define the "Just City," the answer to this question was the philosopher-king, a concept rejected in a long history of struggles to bring arbitrary monarchical behavior under control.

6 As Habermas (1973: 33) has put it, "decisions for . . . political struggle cannot at the outset be justified theoretically and then carried out organizationally." The only "possible justification at this level is consensus aimed at in practical discourse, among participants who, in consciousness of their common interests and their knowledge of circumstances, of the predicable consequences of and secondary consequences, are the only ones who can know what risks they are willing to undergo, and with what expectations." The planner, in short, can bring theoretical and empirical knowledge to bear on the participant's circumstances, but it is the participants themselves through discussion and deliberation who must actually decide which course of action they are willing to undertake.
7 For Healey (1997: 244–5), "a process of deliberate paradigm change" refers to the need to engage "the consciousness of the political and organizational culture of a place." Envisioned as an enlightenment strategy, communicative planning should involve "embedding new cultural conceptions, systems of understanding and systems of meaning" in the world views of social actors.
8 Discursive planning can, for this purpose, borrow from the practices of "transformative learning" worked out by radical educational theorists such as Paulo Freire (Fischer and Mandell 2009).
9 For a detailed discussion of the informal logic of normative discourse based on the work of "ordinary language" philosophers, especially as it applied to policy analysis, see Fischer (1995, 2003).
10 This should not be taken to mean that no one in the planning office thinks about these larger questions. Some surely do, but they seldom express these issues in the work setting (Needleman and Needleman 1974). They are ruled by those who control the agency to be beyond the tasks at hand and thus generally judged to be beside the point, if not inappropriate.
11 Such criticism can range from theoretical naiveté to bureaucratic obfuscation and political mystification.

References

Fainstein, S.S. (2000) "New Directions in Planning Theory," *Urban Affairs Review*, 35: 451–78.

—— (2008) "Spatial Justice and Planning," paper presented at the International Conference on "*Justice et Injustice Spatial*," University of Paris X, Nanterre, March 12–14: 21–22.

Feldman, M. (1997) "Can We Talk? Interpretive Planning Theory as Comedy," *Planning Theory*, 17: 43–64.

Fischer, F. (1995) *Evaluating Public Policy*, Belmont, CA: Wadsworth.

—— (2003) *Reframing Public Policy: Discursive Politics and Deliberative Practices*, Oxford: Oxford University Press.

—— (2009) *Democracy and Expertise: Reorienting Policy Inquiry*, Oxford: Oxford University Press.

Fischer, F. and Forester, J. (eds) (1993) *The Argumentative Turn in Policy Analysis and Planning*, Durham, NC: Duke University Press.

Fischer, F. and Mandell, A. (2009) "Michael Polanyi's Republic of Science: The Tacit Dimension." *Science as Culture*, (18)1.

Flyvbjerg, B. (1996) "The Dark Side of Planning: Rationality and Realrationalitaet," in Mandelbaum, S.J., Mazza, L., and Burchell, R.W. (eds) *Explorations in Planning Theory*, New Brunswick, NJ: Rutgers University Press.

Flyvbjerg, B. (2000) *Making Social Science Matter*, Cambridge: Cambridge University Press.
—— (1999) *The Deliberative Practitioner: Encouraging Participatory Planning Processes*, Cambridge: Cambridge University Press.
Gans, H.J. (1974) *More Equality*, New York: Vintage.
Habermas, J. (1973) *Legitimation Crisis*, Boston: Beacon Press.
—— (1987) *The Theory of Communicative Action*, 2 vols, Cambridge, MA: Polity.
Hajer, M.A. (1995) *The Politics of Environmental Discourse*, Oxford: Oxford University Press.
Hajer, M.A. and Wagenaar, H. (eds.) (2003) *Deliberative Policy Analysis: Understanding Governance in the Network Society*, Cambridge: Cambridge University Press.
Healey, P. (1997) *Collaborative Planning*, London: Macmillan.
Hoch, C. (1994) *What Planners Do: Power, Politics & Persuasion*, Chicago: Planners Press.
—— (1997) "Planning Theorists Taking the Interpretive Turn Need not Travel the Political Economy Highway," *Planning Theory*, 17: 13–37.
—— (2007) "Pragmatic Communicative Action Theory," *Journal of Planning Education and Research*, 26(3): 272–83.
Huxley, M. and Yiftachel, O. (2000) "New Paradigm or Old Myopia? Unsettling the Communicative Turn in Planning Theory," *Journal of Planning Education and Research*, 19: 333–42.
Innes, J.E. (1995) "Planning Theory's Emerging Paradigm: Communicative Action and Interactive Practice," *Journal of Planning Education and Research*, 14(3): 183–89.
Lauria, M. and Whelan, R. (1995) "Planning Theory and Political Economy: The Need for Reintegration," *Planning Theory*, 14: 8–33.
Lindblom, C.E. (1990) *Inquiry and Change: The Troubled Attempt to Understand and Shape Society*, New Haven, CT: Yale University Press.
Lindblom, C.E. and Cohen, D. (1979) *Usable Knowledge: Social Science and Social Problem Solving*, New Haven, CT: Yale University Press.
Majone, G. (1989) *Evidence, Argument, and Persuasion in the Policy Process*, New Haven: Yale University Press.
Mazza, L. (1995) "Technical Knowledge, Practical Reason and the Planner's Responsibility," *Town Planning Review*, 66: 389–409.
Needleman, C. and Needleman, M. (1974) *Guerillas in the Bureaucracy*, New York: Wiley.
Okun, A.M. (1975) *Equality and Efficiency: The Big–Tradeoff*, Washington, D.C.: Brookings Institution.
Ploger, J. (2001) "Public Participation and the Art of Governance," *Environment and Planning B. Planning and Design*, 28: 219–41.
Rawls, J. (1971) A Theory of Justice, Boston: Belknap.
Rawls, J. and Kelly, E. (2001) *Justice as Fairness: A Restatement*, Cambridge: Harvard University Press.
Sager, T. (1994) *Communicative Planning Theory*, Avebury: Aldershot.
Sandercock, L. (1997) *Towards Cosmopolis: Planning for Multicultural Cities*, New York: Wiley.
Shapiro, M. (1981) *Language and Political Understanding: The Politics of Discursive Practice*, New Haven, CT: Yale University Press.
Throgmorton, J.A. (1996) *Planning as Persuasive Storytelling: The Rhetorical Construction of Chicago's Electric Future*, Chicago: University of Chicago Press.

—— (2003) "Planning as Persuasive Storytelling in a Global-Scale Web of Relationships," *Planning Theory*, 2(2): 125–51.

Yiftachel, O. (1998) "Planning and Social Control: Exploring the Dark Side," *Journal of Planning Literature*, 12: 395–406.

Yiftachel, O. and Huxley, M. (2000) "Debating Dominance and Relevance: Notes on the Communicative Turn in Planning Theory," *International Journal of Urban and Regional Research*, 24(4): 907–13.

4 Justice and the spatial imagination[1]

Mustafa Dikeç

In an article published almost two decades ago, H.G. Pirie wrote:

> It would be a pity indeed if the busyness of political philosophers was to go completely unnoticed by spatial theorists and applied researchers. Equally, it would be a pity if this essay were to stand alone as a review of implications of that busyness.
>
> (Pirie 1983: 472)

In that article, entitled "On Spatial Justice," Pirie reflected "on the desirability and possibility of fashioning a concept of spatial justice from notions of social justice and territorial social justice" (1983: 465). This essay offers yet another reflection on the notion of justice as it relates to space and spatiality, to point to the ways in which various forms of injustice are manifest in the very process of spatialization, and the ways in which an increased awareness of the dialectical relationship between (in)justice and spatiality could make space a site of politics in fighting against injustice. As will become clear further through the text, the conceptualizations of both justice and space differ from the ways Pirie once saw them.

GEOGRAPHICAL ENCOUNTERS

Geography's engagement with social justice started with an exclusive concern for distribution in a Cartesian space (Davies 1968); continued with an emphasis on production, the workings of the capitalist city, and uneven geographies of capitalism (Harvey 1973, 1996); shifted focus with a celebration of identity and difference, and an emphasis not on justice and consensus, but on injustice and dissensus (Harvey 1992; Merrifield and Swyngedouw 1997; Merrifield 1997; Gleeson 1998); and ended up, for the moment, with a problematization of the preoccupation with difference, and a search for universal bonds of solidarity (Smith 1997, 2000). There have been few studies, however, that explicitly employed a notion of "spatial justice" (Flusty 1994; Soja 2000). Pirie's spatial sensibilities were certainly intriguing. His conceptualization, however, deserved a reservation. He, admittedly, treated space "in the [then-]familiar way as some kind of container, as an entity or physical expression made up of individual locations and their distance relations" (1983: 471). It is a pity that Pirie stopped at what seems to be a perfect starting point:

> This notion of space is not inviolate. It may also be conceived of as a social creation—as a structure created by society and not merely as a context for society . . . Conceptualizing spatial justice in terms of a view of *space as process*, and perhaps in terms of *radical notions of justice*, stands as an exacting challenge and, not unlikely, as the single occasion there might be for requiring and constructing a concept of spatial justice . . . In spite of the challenge of spatial fetishism, and in spite of the radical assault on liberal distributive concerns, it would be worthwhile investigating the possibility of matching justice to notions of socially constructed space.
>
> (1983: 471–2; emphases added):

What follows, then, is an endeavor that undertakes this enticing suggestion.

To begin, I draw on Iris Marion Young's (1999) definition of injustice, but diverge from her position in that I do not do so in the name of "difference." Rather, I turn to Balibar's notion of égaliberté (equality-freedom) as the ethico-political bond that would inform emancipatory movements, not in the name of a certain particularity but, rather, of an enactment of equality and freedom. Égaliberté provides the "universal" bond of solidarity, and more on the notion may be found in the concluding section.

Underlying these premises, and indeed the whole endeavor, is the conviction that approaches to and principles of justice are time and space specific. A close relationship, therefore, between socio-spatial specificities and conceptions of justice is assumed. I should like to believe that a sensitivity to the spatial dimension of justice may be developed—especially in societies where the injustices of spatial dynamics are exposed and largely recognized—to guide emancipatory movements to suppress domination and oppression in and through space. The city seems to provide a fertile ground for such a prospect.

URBAN SENSIBILITIES

The impulses of the 1960s and 1970s were influential in bringing the "urban question" into the political agenda. Two major reasons for the development of an urban sensibility in this period may be defined. First, there was a growing reaction to the functional approach to the city, and to the growing emphasis on the city's exchange value to the detriment of its use value. Second, as Bertho (1999) states, there was a search for alternative social movements to the workers' movement, for a political mobilizing force that was more spatial and urban in nature than the traditional mobilizations based on work. These reasons, of course, may less confidently be stated outside the French political culture. However, it was in this context and for these concerns that Henri Lefebvre conceived the notion of right to the city.

"The right to the city," Lefebvre (1993: 435) writes, "cannot be considered a simple visiting right or a return to the traditional city. It can only be formulated as the *right to urban life*, in a transformed and renewed form." It is not simply the right of property owners, but of all who live in the city. The right to the city, therefore, does not imply a "clean" and quaint city where the "good citizens" mingle on its streets, crowding its beautiful parks, and living there happily ever after. As Lefebvre (1996: 195) argues, "[I]t does not abolish confrontations and struggles. On the contrary!" This, of course, is an unsurprising claim given Lefebvre's conception of the city as the place of simultaneity and encounter, and of rights to be given content through struggle. And he is perfectly sensible to ask the question (1993: 428): "are these not specific urban needs? Is there not also the need for a time for such encounters, such exchanges?" There is, I think, a double message in these questions; an appeal and a critique. The appeal points to the need to (re)claim the right to the city. The critique, on the other hand, stems from Lefebvre's denouncement of the welfare capitalist society as the "bureaucratic society of organized consumption" where needs are created and institutionalized, where exchange value of urban space is prioritized over its use value. It was a critique, in the context of the 1960s and 1970s when functionalist and technocratic urbanization processes were under severe critique for eradicating urbanity, and depriving urban dwellers of places of social encounter through the rational ordering of urban space. This critique, in a sense, is a critique of the abstraction of rights from the city which, I believe, was the motivation for Lefebvre to advance a notion of right to the city. Let me try to clarify the point.

A common theme flowing through all the writings of Lefebvre on the city is a critique of the contemporary city, which had started to deteriorate in the nineteenth century with the development and deployment of industrial capitalism and the rise of the modern nation-state. In the twentieth century, as capitalism dominated all areas of social life, society was turned into a "bureaucratic society of controlled consumption." Not only was consumption controlled, but the spaces of the society and their production as well, with the city being the site where these powers were at work more intensely. In that sense, Lefebvre's attempt can be seen as a political project to rescue individuals from oppressive and homogenizing processes by asserting their right to the city. "The individual does not disappear in the midst of the social effects caused by the pressures of the masses, but is instead affirmed," he would write. "Certain *rights* come to light" (1993: 435).

What, then, happened in the nineteenth and twentieth centuries, besides the development and deployment of capitalism, which would lead one to point to the need to reassert rights to the city? And what kind of a right is that;

why is Lefebvre so careful to note that it "cannot be considered a simple visiting right?" Let me start with the latter question. Such a conception would bring him too close to the argument Kant advanced in his famous text on "Perpetual Peace" concerning world citizenship and universal hospitality. "The stranger, states Kant (1970), cannot claim a *right of residence* but a *right of visit*." The stranger, upon arrival, also enjoys another right, that of hospitality, which suggests that they not be treated as an enemy by the host of the territory in question. There is no room for the stranger to claim a right, but simply to enjoy a right to visit or pass through. Besides, this right is conceived as an interstate conditionality, and thus, there is no hospitality for those who do not possess citizenship status in one of the signatory states.[2]

The answer to the first question, on the other hand, may be formulated in relation to changing conceptions of citizenship, a notion that has currently been challenged, as, for example, Wihtol de Wenden (1992) suggests, by two fundamental elements of political context and social tissue: immigration and urbanity. Citizenship originated as rights to the city in a spatial sense, which are now abstracted from the city, from their spatial origins. And this detachment coincides with the periods of which Lefebvre is so critical, and which eventually leads him to proclaim the need to reclaim rights to the city. The city, Isin (1999: 165) argues, "has not only been a foreground or a background to struggles for group rights but also a battleground to claim those rights." These struggles were pursued as "claiming rights to the city as a space of politics" up until the seventeenth century, where citizenship and struggles were redirected to center on the state and, eventually, the modern nation-state since the nineteenth century. In this sense, Lefebvre's notion of the right to the city is a call to advance an urban spatial approach to political struggles with the participation of all those who inhabit the city without discrimination.

> The right to the city, complemented by the right to difference and the right to information, should modify, concretize and make more practical the rights of the citizen as an urban dweller (*citadin*) and user of multiple services. It would affirm, on the one hand, the right of users to make known their ideas on the space and time of their activities in the urban area; it would also cover the right to the use of the center, a privileged place, instead of being dispersed and stuck into ghettos (for workers, immigrants, the 'marginal' and even for the 'privileged').
>
> (Lefebvre 1986: 170; translation from 1996: 34)

The right to the city implies not only the participation of the urban citizens in urban social life but, more importantly, their active participation in the political life, management, and administration of the city. The achievement of these rights, Lefebvre (1986) states, supposes the transformation of the society, of time and space. It is the urban political life to be changed, not the city *per se*. In other words, the right to the city entails not at all a right to be distributed from above to individuals, but a way of actively and collectively relating to the political life of the city. The urban would then consist of "a civil society founded not on abstractions but on space and time, as '*lived*'" (1986: 173; emphasis added). The right to the city, therefore, is not simply a participatory right but, more importantly, an *enabling right*, to be defined and refined through political struggle. It is not only a right to urban space, but to a political space as well, constituting the city as a space of politics. Urban citizenship, in this sense, does not refer to a legal status, but to a form of identification with the city, to a political identity. The construction of this identity through political struggle is enabled by another right—the right to difference.

The right to difference is complementary to the right to the city. Lefebvre was very clear on this notion, although it somehow ended up with an exclusive focus on difference as particularity. The right to be different, he wrote, is "the right not to be classified forcibly into categories which have been determined by the necessarily homogenizing powers" (1976a: 35). His emphasis was on the "be" of the "right to be different," not particularly on the "different," narrowly interpreted as particularity. Therefore, the connotations were disagreement and contestation (and eventually *differ*ing) resulting from a right to "not to subscribe to," from a "right to *be* different." A better translation, perhaps, would be "the right to resist/struggle." The difference between particularities and differences, and the dynamic relation between them were part of Lefebvre's "differentialist" project. He was critical of the reduction of the latter to the former, and the movement from the former to the latter was the moment of differing, achieved only through political struggle (Lefebvre 1981). Solidarity, therefore, was not built merely around particularities, as Young's (1990) scheme would suggest, for example, but on the will and capabilities to differ. Criticizing the marginalization of certain groups through identities imposed upon them or through the use of identities that they associate themselves with, and then trying to develop a "politics of identity" in the name of those "differences" to resist such processes is simply to accept and remain trapped in the already established categories.

For Lefebvre particularities existed naturally, but then became difference in the modern world. It is the concept of difference that is created by distanced reflection, and the illusory difference that individuals feel which, eventually, makes them indifferent. He writes:

> [D]ifference in act differs . . . from the difference merely thought or reflected. The thought and not-lived difference of philosophers and logicians is in opposition with the non-thought and lived difference . . . [I]t cannot be reduced to banalized re-presentations: originality, diversity, variety, distinction, etc.
>
> (1970a: 65–66)

The right to difference, therefore, is the basis and source of other concrete rights which could "be fully affirmed only beyond the written and the prescribed, in a practice recognized as the basis of social relations" (1970a: 45). "Differentialism," as he states at the conclusion to his manifesto, "is about living. Not thinking but 'being' differently" (1970a: 186).

Therefore, the rights that Lefebvre conceptualizes are established through lived experience and social relationships, and once established, they would lead to new ways of life, new social relations, and possibilities to differ—even concerning the established rights themselves. In this sense, rights, as conceived by Lefebvre, become

> more of a claim upon than a possession held against the world. It becomes a claim upon society for the resources necessary to meet the basic needs and interests of members rather than a kind of property some possess and others do not . . . [I]n terms of rights to the city and rights to political participation, right becomes conceived as an aspect of social relatedness rather than as an inherent and natural property of individuals.
>
> (Holston and Appadurai 1996: 197)

In claiming these rights, a notion of spatial justice might serve as a mobilizing discourse through the cultivation of a spatial sensibility toward injustice, and a spatial culture to fight against it. It might also serve as an ethico-political imperative to avoid the abusive interpretations of these rights.

SPATIAL JUSTICE: A CONCEPTUALIZATION

I begin with three exemplary cases to point to the role of spatialization in the maintenance and manifestation of injustice. The first one is Harvey's account of the Imperial Foods plant fire in Hamlet. In comparing the incident in Hamlet with the one in New York, to point to the lack of political response in the former, Harvey writes:

> A similar event in a relatively remote rural setting posed immediate logistical problems for massive on-the-spot political responses (such as the protest demonstration on Broadway), illustrating the effectiveness of capitalist *strategies of geographical dispersal* away from politicized central city locations as a means of labor control.
>
> (1996: 340; emphasis added)

This was, perhaps, the great lesson that industrial capitalists had learned from the grand strike of July 1969 in Turin, organized by the united syndicates and political parties of the left against increased living costs. The 600,000 workers united in Turin were not only exploited in the factory, but dominated in their city as well—both consequences of the logic of the capitalist mode of accumulation. Exploitation was produced and reproduced by social relations of power established under the capitalist production system, while spatial domination was produced *and* reproduced by the spatial logic of capitalism, contributing further to the domination of a certain group of the population. In this sense, the Turin strike was "one of the first movements for the right to the city against the pattern in which the city was developed under the pressure of speculation, motivated by the logic of maximum profit" (Novelli, cited in Lojkine 1977: 335). Injustice in the factory was exploitation. Injustice in the city was the domination of urban space, pushing the workers away from the city where rent was no longer affordable. Injustice was at once socially and spatially manifest, and above all, was produced not only socially but spatially as well.

The case for the injustice of spatialization was made neither for Hamlet nor Turin. The case, however, was made in the Bus Riders Union (BRU) in Los Angeles, if not explicitly in these terms. The argument was that the Metropolitan Transit Authority's (MTA) transit policies and investment schemes were discriminative against a particular population of transit-dependent bus-riders. The case was brought to court (*Labor/Community Strategy Center v. Los Angeles Metropolitan Transit Authority*) as a class action suit on behalf of 350,000 bus riders. It was resolved in 1996 through a Consent Decree, forcing the MTA to reconsider its policies, resulting in a temporary stop of the construction of the planned fixed-rail transit system. What was questioned in the case was the spatial pattern imposed by the MTA on the transit-dependent poor working populations of Los Angeles. In this sense, the BRU

> can be seen as opening traditional notions of civil rights to a more specifically spatial politics revolving around new visions of democratic citizenship and the rights to the city, the rights—and responsibilities—of all urban dwellers to participate effectively in the social production of their lived cityscapes.
>
> (Soja 2000: 257–8)

Moreover, Soja argues, the BRU case represented an attentiveness to see injustice in its spatial dimension, signaling, perhaps, the formation of spatially informed practices and politics—what he calls "spatial justice":

> I do not mean to substitute spatial justice for the more familiar notion of social justice, but rather to bring out more clearly the potentially powerful yet often obscured spatiality of all aspects of social life and to open up in this spatialized sociality (and historicality) more effective ways to change the world for the better through spatially conscious practices and politics.
>
> (Soja 2000: 352)

I read these examples as illustrative vignettes that denote both the "spatiality of injustice"—from physical/locational aspects to more abstract spaces of social and economic relationships that sustain the production of injustice—and "injustice of spatiality"—the elimination (i.e., in the Hamlet case) of the possibilities for the formation of political responses (i.e., in the Turin and Los Angeles cases).

In the dialectical formulation of the *spatiality of injustice* and the *injustice of spatiality*, the former notion implies that justice has a spatial dimension to it, and therefore, that a spatial perspective might be used to discern injustice *in space* (which, of course, can effectively be captured by an analysis of distribution patterns, as the BRU case exemplifies). The latter, on the other hand, implies existing structures in their capacities to produce and reproduce injustice *through space*. It is, compared to the former, more dynamic and process-oriented. Such a conceptualization implies two essential points. First, analysis should not be based on the thing under consideration *per se*, but also on the components of it. Second, form and process are inseparable and should be considered together. How, then, may such an approach be related to space? As Lefebvre puts it:

> The dialectic is back on the agenda. But it is no longer Marx's dialectic, just as Marx's was no longer Hegel's . . . The dialectic today no longer clings to historicity and historical time, or to a temporal mechanism such as 'thesis-antithesis-synthesis' or 'affirmation-negation-negation of the negation' . . . To recognize space, to recognize what 'takes place' there and what it is used for is to resume the dialectic; analysis will reveal the contradictions of space.
>
> (Lefebvre 1976a: 14 and 17).

> (Social) space is not a thing among other things, nor a product among other products . . . It is the outcome of a sequence and set of operations, and thus cannot be reduced to the rank of a simple object . . . Itself the outcome of past actions, social space is what permits fresh actions to occur, while suggesting others and prohibiting yet others.
>
> (Lefebvre 1991: 73)

Therefore, the emphasis is not on space *per se*, but the processes that produce space and, at the same time, the implications of these produced spaces on the dynamic processes of social, economic, and political relations. The basic features of the dialectical formulation I propose to consider in the relationship between injustice and spatiality are, therefore, as follows:

a. focusing on spatiality as a process; as a producer and reproducer of, at the same time being produced and reproduced by, relatively stable structures (permanences);
b. recognizing the interrelatedness of injustice and spatiality as producing, reproducing, and sustaining each other through a mediation of larger permanences that give rise to both of them.

In this sense, the notion of spatial justice is a critique of systematic exclusion, domination, and oppression; a critique aimed at cultivating new sensibilities that would animate actions towards injustice embedded in space and spatial dynamics. The aim is to explicate an ideological discourse on the spatiality of (in)justice, informed by the two notions of right to the city and right to difference/resist, into which such emergent movements as the BRU may insert themselves. Under a larger, but not a totalizing, conception of justice, it helps to assess the processes that could be the sources and resources of injustice in their specific contexts, for similar processes may produce different consequences in different contexts. Focusing on processes in their relationships and implications would be a reformulated dialectical approach as opposed to focusing on fixed forms or distributive consequences under a universal stencil of justice. Such an approach helps to discern not whether a particular event is just or unjust, but rather, to explore the dynamic processes of social, spatial, economic, and political formations in order to see if they operate in such a way to produce and reproduce dominant and oppressive permanences which would be considered as being unjust. In a spatial sense, domination as a form of injustice manifests itself in space; most visibly in the built environment but also in other various forms of less visible (or not visible at all) spaces of flows, distributions, networks, and institutions. More importantly, space and the processes of spatialization play a major role not only in the production of the conditions of domination, but also in their reproduction and survival as an indispensable manipulative tool for the existing mode of production (i.e., capitalism).

How, then, do the three notions I have employed thus far—spatial justice, right to the city, right to difference—come together as part of an emancipatory politics? In what follows, I shall offer a "triad" that brings together these notions, and defines the parameters of a spatially informed emancipatory politics.

The triad consists of three notions: spatial dialectics of injustice (spatiality of injustice and injustice of spatiality), right to the city, and right to difference. I articulate these notions as the parameters of an "ideal of égaliberté" (equality-freedom); that is, the suppression of domination and repression, the achievement of which would imply a moment of emancipation. The argument is that Lefebvre's two rights, which have been the subject of much intellectual effort aimed at emancipatory politics, provide a better framework when articulated with a "spatial dialectics of injustice." This, indeed, seems to be a necessary underpinning in order to resolve the paradox rising from decidedly individualistic versus collective interpretations of these rights.[3] Such an approach is informed not only by the premise that spatial dynamics play an essential role in the production and reproduction of injustice, but also by the premise that they permit, as well as prohibit, the formation of rights claims, and the ways and extent in which rights are put into action, or practiced.

The triad provides a common lexicon[4] and conceptual apparatus, which then could cultivate an "ethico-political bond" (Mouffe 1992: 231), or an "ethics of political solidarity" (Harvey 1996: 360) to inform emancipatory movements of those,[5] to name a few, "trapped in space" (Harvey 1989), "chained to a place" (Bourdieu 1999), "disabled by the social production of space" (Gleeson 1998), excluded by urban entrepreneurialism (Hall and Hubbard 1996), or expelled through urban renewal projects (Leroux 2001). The bond is ethical in the sense that it is nurtured through social relatedness, rather than assumed as an ontological given. Such a conception is necessary to imply that forms of morally defensible practices must be socially negotiated, through engagement, rather than being manipulated by the anxieties of dominant groups in the city. And it is political in the sense that it entails antagonism and contestation, and not always an effortless reconciliation, for the simple reason that the very production of space is decidedly political (see, for example, Lefebvre 1976b). As for the way in which the individual components of the triad relate to one another, the notion of spatial (in)justice sets the parameters by which the right to the city may be assessed, violations of which are resisted through a right to difference. And there is good reason to believe that such a spatial sensitivity to (in)justice may be developed, especially in societies where the spatial dynamics of social exclusion are widely recognized. Miller's thesis endorses this likelihood, according to which people

> hold conceptions of social justice as part of more general views of society, and that they acquire these views through their experience of living in actual societies with definite structures and embodying particular kinds of interpersonal relationship.
>
> (Miller 1976: 342)

CONCLUSIONS: POLITICIZING THE URBAN SPATIAL

> Space is a doubt: I have constantly to mark it, to designate it. It's never mine, never given to me, I have to *conquer* it.
>
> Georges Perec (1997: 91)

> In the case of ideologies of what is good and right it may be space rather than time that is crucial. Something may be good and just everywhere, somewhere, here or elsewhere.
>
> Goren Therborn (cited in Cresswell 1996: 3)

"The limits of just and unjust," Voltaire once wrote, "are very difficult to set down; like the middle state between health and illness, between the appropriateness and inappropriateness of things, between the false and true, is difficult to mark." It would, similarly, be very difficult, and hardly desirable, to discern *the* line between justice and injustice as they relate to spatial practices. For this reason, perhaps, it is best to conceive the notion of spatial justice, with the rights to the city and difference, in relation to a "universality of an ideal," as exemplified by Étienne Balibar's proposition of égaliberté (equality-freedom).

Égaliberté signifies the unconditional "differential" push in the collective process of struggle for the suppression of discrimination and repression. Balibar (1997) defines "equality" as non-discrimination, and "freedom" as non-constraint (non-repression), both of which remain open to diverse determinations depending on the circumstances. Eliminating or simply fighting against discrimination necessarily implies the elimination or the fight against repression and vice versa (the impossibility of freedom without equality and of equality without freedom)—hence the notion of égaliberté. The joint suppression of discrimination and repression implies emancipation.

Balibar's proposition, as a politics of emancipation, has important implications for the current project for two major reasons. First, égaliberté, as an ideal of non-discrimination and non-repression, remains as an indestructible resource of insurrection against the existing order, reborn and re-experienced in diverse situations, places, groups, etc. It is achieved through a *collective process*, through a political struggle "against oppression, social hierarchies and inequalities." Equality and freedom, in this collective process of the suppression at once of discrimination and repression, therefore, "can never be granted, distributed among individuals, they can only be *conquered*" (1997: 446; emphasis added). This has close resonance with Lefebvre's "differentialism" and the rights to the city and difference, for both rights are constantly redefined and acquired through political struggle, and they are not procedural and normative in nature. The ideal, then, is to conquer, through a collective process, the spaces (right to the city) and means (right to difference) of this struggle. In this sense, the resonance with égaliberté would be thus: the right of all city habitants to participate in the political life of the city (fighting against discrimination), and their right to political struggle of resistance (fighting against repression). What is implied is not simply a political movement, but a transformation of the political itself. What is implied is a "*right to politics*," meaning that "no one can ever be emancipated from outside or from above, but only by his/her own action and its collectivization" (Balibar 1997: 446). Rancière (1995: 70), in this sense, was perfectly sensible to state that "the question is not only 'How are we to face a political problem?' but 'How are we to reinvent politics?' "

The second implication of Balibar's *égaliberté* has to do with the notion's—as a universality of an ideal—transindividual nature. This does not only refer to a collective process of struggle against discrimination and repression, but also to its universal symbolic dimension as a politics of emancipation, which means that "it does not depend on the extension of its influence or of its popularity" (1997: 447). The struggle, therefore, is not aimed at defending "the rights of a particular group in the name of that particularity itself, but [at] proclaiming that the discrimination or exclusion that strikes that particular group represents a negation of humanity as such" (1997: 453). What is changed, at the end of the day, is the very political itself: The struggle does not delineate an identity-defined community but a solidarity in the collective process of the elimination of discrimination and repression, in achieving égaliberté, which always remains as the element of subversion of the existing order.

Balibar also has in mind a third notion of equality, neither distributive nor participatory, but an openly *civic* one (a new ethics and politics). Informed by such a conception of equality, he argues (1991: 66), "talking about the right to the city would be a way of indicating that the city becomes as such a *polis*, a political collectivity, a place where public interest is defined and realized." Isn't this what Lefebvre also had in mind with his notion of right to the city complemented by the right to difference? Such a perspective above all implies a conception of the city that goes beyond an administrative entity, and makes it the place in and over which the terms of the right or, better still, the nature of relationships between state, society and its space are negotiated.

Right to the city implies not only a formulation of certain rights and the cultivation of the political among city habitants, but also a reconsideration of the spatial dynamics that make the city. It, therefore, should not be conceived merely as a practice of claiming and asserting rights, but should also be conceptualized in a way so that it puts on the agenda the *dynamics* (e.g., property markets) and *principles* (e.g., urban policy, land use policy, planning laws) of the ways in which social relations are spatialized in the city. The triad I propose, in this sense, is an attempt to address this issue via a notion of spatial justice that calls into question the dynamics and principles that manifest themselves spatially in the city. Besides, it seems necessary, should the Lefebvrian notion be used, to conceptualize a notion of spatial justice in order to be able to distinguish between appropriation and domination of urban space in the name of a right to the city.

Neil Smith's (1997: 134) example of the claim of and the counter-claim against homeless people, *both* of which appeal to the same source for justification, exposes "the doubleness inherent in the system of abstract justice based on individual rights." The former poses the following question: "why do we as homeless individuals have no right to housing?" The latter, against this, and using the same source of appeal, presses another question: "don't I have a right to live without homeless people messing up the neighborhood?" The first question, as Smith states, is "over and against individualism" whereas the second stands for a "reasserted individualism." This is a remarkable example, for it depicts the perils of a right to the city discourse interpreted in the liberal framework of individual rights. The notion of the right to the city implies not only a spatial change, but a societal one as well, for the ways in which the notion is conceived and justified depends largely on the very society itself, its political culture included. In other words, the right to the city might vary drastically depending on the society in question, especially if the demos is marked with clear and rigid demarcations. In the Greek polis, for example, the right to the city arguably was perfectly practiced except by women and slaves. To take a recent example, there was nothing to prevent former New York City Mayor Rudolph Giuliani from advancing a "right to the city" (for some but not for others), and enforcing Zero Tolerance policies against the homeless in its name unless a new societal ethics were to be cultivated by the living togetherness of *all* city inhabitants. How, then, could one overcome this paradox, and conceive of the right to the city (and to difference as well) as part of an emancipatory politics? Through the "heterology of emancipation," I would argue following Jacques Rancière.

"[T]he logic of political subjectivization, of emancipation," writes Rancière; "is a heterology, a logic of the other . . . it is never the simple assertion of an identity" (1995: 68). Even the workers' movement was not an assertion of identity. "Proletarii," in Latin, meant "prolific people"—"people who make children, who merely live and reproduce without a name, without being counted as part of the symbolic order of the city." *Proletarians*, therefore, did not imply a particular identity; it, rather, was "the name of anyone, the name of the outcast: those who do not belong to the order of castes, indeed, those who are involved in undoing this order" (Rancière 1995: 67). Therefore, their movement, as one of emancipation, did not imply a selfish reassertion of individuality, but an enactment of equality and freedom:

> One can object that the idea of emancipation is historically related to the idea of the self in the formula of "self-emancipation of workers." But the first motto of any self-emancipation movement is always the struggle against "selfishness." This is not only a moral statement (for instance, the dedication of the individual to the militant community); it is also a logical one: the politics of emancipation is the politics of the self as an other, or, in Greek terms, a "heteron." The logic of emancipation is a heterology.
>
> (Rancière 1995: 65)

Seen in this way, égaliberté, as an enactment of equality and freedom, provides the universal that transcends particular identifications. The triad I have offered, in this sense, defines the spatial parameters of a struggle against discrimination and repression. The formation of a sensibility to the spatial dimension of injustice could be considered as a transcendence of tensions rising from particular group interests in the city (segregation, socio-spatial exclusion, etc.) through a universality as an ideal, through the development of a consciousness nurtured by living together and sharing space, through a (civic) ideal of égaliberté. If the city remains a site of structural socio-spatial exclusion, where domination is legitimated, not only such a consciousness remains undeveloped, but, worse still, misdeveloped; that is, giving rise to socio-spatial exclusion in and against the city. All this is not to imply that the city, through its evolution, has not been the site of exclusion and segregation. Neglecting this fact would be a nostalgic fallacy. Legitimating, even fostering, this fact, however, would be a fatal flaw, engendering Zero Tolerance type of policies. How, then, would the formation of such a consciousness contribute to the struggle against injustice? How may a spatially informed and sensible politics of resistance use space in the fight against injustice? How, in other words, could a spatial content be given to such a politics of emancipation?

Four approaches, following Garber (2000: 267–9), may be distinguished, which, of course, are not mutually exclusive, and which could also work against emancipatory (égaliberté) movements. A politics of resistance could use urban space in ways so that: first, people act *from* space, politically mobilizing from the material conditions of their space, and seeking alternative spatializations (for example, the Paris Commune, Los Angeles riots, the Turin strike, the BRU case); second, people act *on* space, to appropriate or to dominate it with a group identity (for example, appropriation of the Tompkins Square Park in New York, but also the formation of gated communities); third, people act *in* space, taking it to the streets for debates, displays, protests or violence (for example, Paris Commune, May 1968, Los Angeles riots, the Turin strike); and finally, people *make* space, creating the conditions to expand the public political involvement through the linking of metaphorical space and politics (for example, in Harvey's account above, the formation of a political response in New York, but not in Hamlet, the organization of the BRU case). Although space most often appears as a means of control and domination, it, and for this very reason, also carries the seeds of significant resistance (Cresswell 1996).

What, then, are the substantive implications of the notion of spatial justice? Two mutually constitutive prospects may be discerned. First, it might provide a conceptual apparatus that could be given normative content to guide the actual production of urban space. Second, its discursive development and deployment might inform emancipatory politics willing to confront spatial dynamics that produce and reproduce various forms of injustice. Both prospects, of course, will depend on the society in question, and the meanings attached to principles of justice. Justice, after all, is a contingent "reification of commensurability" (Dimock 1997: 6), not an ontological given.

NOTES

1 This chapter consists substantially of material previously published in: Dikeç (2001) "Justice and the Spatial Imagination," *Environment and Planning A*, 33: 1735–1805. Sections of this piece are reprinted here with the permission of the publisher, Pion Ltd. I am grateful to Cuz Potter, Johannes Novy, Ingrid Olivo and James Connolly for their careful editing.
2 An elaboration of the notion of hospitality, pointing to the limits and perils of its Kantian version, may be found in Dikeç (2002).
3 "A city center for all" is an association in Marseille, currently fighting against speculative forces generated by urban renewal projects, which force out the immigrant population from the city center. See Leroux (2001).

4 Michael Keith (1997) employs a similar notion—"vocabulary of resistance"—in his work on young Bengalis in London's East End, and shows how such a common lexicon serves as a bond in their struggle against racism.
5 This, of course, must not be interpreted as implying merely "those who are affected," but be seen in the light of the notion of égaliberté. In this sense, these movements represent a resistance against the negation of socially enshrined principles (in this case, of the suppression of discrimination and exclusion), against the negation of humanity as such (Balibar 1997). This is the universal bond which coalesces not only between those who are affected, but all who wish to negate the negation of humanity.

REFERENCES

Balibar, E. (1991) Comments in the minutes of seminar "Loi d'orientation pour la ville: seminair rechercheurs decideurs," *Recherches* 20.

—— (1997) *La crainte des masses. Politique et philosophie avant et après Marx*, Paris: Galilée.

Bertho, A. (1999) *Contre l'État, la politique*, Paris: La Dispute.

Bourdieu, P. (1999) "Site effects," in Bourdieu, P. *et al.* (eds) *The Weight of the World: Social Suffering in Contemporary Society*, trans. by Ferguson, P.P. *et al.* [*La misère du monde*, 1993, Editions du Seuil, Paris], Palo Alto: Stanford University Press; 123–9.

Cresswell, T. (1996) *In Place/Out of Place: Geography, Ideology, and Transgression*, Minneapolis: University of Minnesota Press.

Davies, B. (1968) *Social Needs and Resources in Local Services*, London: Michael Joseph.

Dikeç, M. (2002) "Pera peras poros: longings for spaces of hospitality," *Theory, Culture & Society*, 19 (1–2): 227–247.

Dimock, W.C. (1997) *Residues of Justice: Literature, Law, Philosophy*, Berkeley and Los Angeles: University of California Press.

Flusty, S. (1994) *Building Paranoia: The Proliferation of Interdictory Space and The Erosion of Spatial Justice*, West Hollywood, CA: LA Forum for Architecture and Urban Design.

Garber, J.A. (2000) "The city as a heroic public sphere," in Isin, E. (ed.) *Democracy, Citizenship and the Global City*, London and New York: Routledge; pp. 257–274.

Gleeson, B. (1998) "Justice and the disabling city", in Fincher, R. and Jacobs, J.M. (eds) *Cities of Difference*, New York and London: The Guilford Press; pp. 89–119.

Hall, T. and Hubbard, P. (1996) "The entrepreneurial city: new urban politics, new urban geographies?," *Progress in Human Geography*, 20(2): 153–174.

Harvey, D. (1973) *Social Justice and the City*, Baltimore: Johns Hopkins University Press.

—— (1989) *The Urban Experience*, Baltimore: Johns Hopkins University Press.

—— (1992) "Social justice, postmodernism and the city," *International Journal of Urban and Regional Research*, 16(4): 588–601.

—— (1996) *Justice, Nature and the Geography of Difference*, Malden, MA and Oxford: Blackwell.

Holston, J. and Appadurai, A. (1996) "Cities and citizenship," *Public Culture*, 8(2): 187–204.

Isin, E. (1999) "Introduction: cities and citizenship in a global age," *Citizenship Studies* 3(2): 165–172.

Kant, I. (1970 [1795]) "Perpetual peace: a philosophical sketch," in *Kant's Political Writings*, ed. Reiss, H. trans. Nisbet, H.B., Cambridge, UK: Cambridge University Press; pp. 93–130.

Keith, M. (1997) "Street sensibility? Negotiating the political by articulating the spatial," in Merrifield, A. and Swyngedouw, E. (eds) *The Urbanization of Injustice*, New York: New York University Press; pp. 137–160.

Lefebvre, H. (1970) *Le manifeste différentialiste*, Paris Gallimard.

—— (1976a [1973]) *The Survival of Capitalism: Reproduction of the Relations of Production*, trans. Bryant, F., London: Allison & Busby.

—— (1976b) "Reflections on the politics of space," *Antipode*, 8: 30–37.

—— (1981) *Critique de la vie quotidienne III: De la modernité au modernisme (Pour une métaphilosophie du quotidien)*, Paris: L'Arche éditeur.

—— (1986) *Le Retour de la Dialectique*, Paris: Messidor/Éditions sociales.

—— (1991 [1974]) *The Production of Space*, Oxford: Blackwell.

—— (1993 [1968]) "The right to the city," in Ockman, J. (ed.) *Architecture Culture 1943–1968: A Documentary Anthology*, New York Rizzoli; pp. 428–436.

—— (1996) *Writings on Cities*, trans. Kofman, E. and Lebas, E., Cambridge, MA: Blackwell.

Leroux, L. (2001) "A Marseille, les anciens de l'immigration, oubliés de la rénovation urbaine," *Le Monde vendredi*, 22 juin, p. 11.

Lojkine, J. (1977) *Le marxisme, l'état et la question urbaine*, Paris Presses Universitaires de France.

Merrifield, A. (1997) "Social justice and communities of difference: a snapshot from Liverpool," in Merrifield, A. and Swyngedouw, E. (eds) *The Urbanization of Injustice*, New York: New York University Press; pp. 200–222.

Merrifield, A. and Swyngedouw, E. (1997) "Social justice and the urban experience: an introduction," in Merrifield, A. and Swyngedouw, E. (eds) *The Urbanization of Injustice*, New York: New York University Press; pp. 1–17.

Miller, D. (1976 [1979]) *Social Justice*, Oxford: Clarendon Press.

Mouffe, C. (1992) "Democratic citizenship and the political community," in Mouffe, C. (ed.) *Dimensions of Radical Democracy: Pluralism, Citizenship, Community*, London and New York: Verso; pp. 225–239.

Perec, G. (1997 [1974]) *Species of Spaces and Other Pieces*, edited and translated by J. Sturrock, Harmondsworth, Middlesex: Penguin Books.

Pirie, H.G. (1983) "On spatial justice," *Environment and Planning A*, 15: 465–473.

Rancière, J. (1995) "Politics, identification, and subjectivization," in Rajchman, J. (ed.) *The Identity in Question*, New York and London, Routledge; pp. 63–70.

Smith, D.M. (1997) "Back to the good life: towards an enlarged conception of social justice," *Environment and Planning D: Society and Space*, 15: 19–35.

—— (2000) "Social justice revisited," *Environment and Planning A*, 32: 1149–1162.

Smith, N. (1997) "Social justice and the new American urbanism: the revanchist city," in Merrifield, A. and Swyngedouw, E. (eds) *The Urbanization of Injustice*, New York: New York University Press; pp. 117–136.

Soja, E. (2000) *Postmetropolis: Critical Studies of Cities and Regions*, Oxford and Cambridge, MA: Blackwell.

Wenden, W. de (1992) "Question de citoyenneté," *Espaces et sociétés*, 68: 37–45.

Young, I.M. (1990) *Justice and the Politics of Difference*, Princeton, NJ: Princeton University Press.

—— (1999) "Residential segregation and differentiated citizenship," *Citizenship Studies*, 3(2): 237–252.

Part II

What are the limits of the Just City?

Expanding the debate

5 From Justice Planning to Commons Planning

Peter Marcuse

> The law locks up the man or woman
> Who steals a goose from off the common,
> But leaves the greater villain loose
> Who steals the common from the goose.

Just city thinking contributes to strengthening the normative claim of urban planning, which is badly needed in the current period of pragmatism and retreat. Yet the call for distributive justice is a necessary but not sufficient aspect of such a normative pitch. It fails to address the causes of injustice, which are structural and lie in the role of power. The jingle above suggests the problem: justice requires the punishment of the thief, but does not prevent theft nor protect the continued functioning of the common. The common represents a whole system of property rights and production relations, not simply equality of use. It suggests a model of the desired city, which should not be a city with only distributional equity, but one that supports the full development of human capabilities for all. That requires more than Justice Planning; it requires Commons Planning. But the discussion should not be about the definition of imaginary utopias. It should rather be used as a way of raising concretely the structural issues that underlie the creation and exercise of power in social relationships, power that both produces distributional injustices and more broadly inhibits the attainment of a good, or humane, or just city.

JUSTICE PLANNING

The call for a Just City is compelling. The demand for justice is a vital one, with centuries of struggle, of interpretation, and of political concern behind it. It is concrete; it is not a call for another world, but for changes in this world, in one clear direction. The exact definition of justice may indeed be controversial, but the philosophical difficulties do not lessen the importance of the formulation. In the context of the existing city, it does not take a great deal of convincing to make concern for its improvement in the general direction of justice an understandable concern. No one will argue that cities should be unjust and no one will deny that the level of justice in today's cities can be much improved. Highlighting this fact moves the entire discussion forward in an important way; introducing the concept of justice, even without resolving all its implications, is a major step forward, and is immediately useful in fields such as planning and social policy.

When we come to look at the role of concepts of the Just City as they would be used in Justice Planning several problems arise. Justice, at least in the widely accepted sense of distributional justice, is a necessary, but not a sufficient, goal of planning. It supports the remedying of injustices and leads towards the achievement of a city without distributional injustices, but it does not address the structural causes of injustice without additions.

There is a further problem: that of defining carefully what the broader goal should be, which, simply put, is the removal of the structural obstacles to a just distribution of goods and services. Various formulations have been suggested, involving terms such as diversity, liberty, capabilities, human development, and an environment of heart's desire. Such goals can of course be subsumed under the concept of justice, but that merely postpones the problem of their definition and legitimation. Other chapters in this volume discuss several of these formulations and their utility is taken up in the Conclusion. This chapter does not focus on these formulations and allows the use of various vague and general phrases, such as "a better society," "serving the common good," and "broader issues" to avoid clouding the main point, which is to raise the issue of how planning can deal with structural issues.

The limits of Justice Planning

Justice is an intuitively meaningful concept: equality of treatment and equity in result, or simply fairness. It is an approach that a socially conscious view would consider the basis for law in urban planning. Legal approaches to planning issues fundamentally judge planning decisions against standards of justice and define what those standards should be. The law legitimates those standards by reference to legislation as the outcome of democratic and just process, "due process," thus combining both the outcome issue and the process issue. Planning, by contrast, is not directly concerned with individual justice, but rather with social utility: a planner would have no hesitation zoning one of four corners at an intersection for a gas station and prohibiting such a use at the other three, while a traditional legal view would demand justification for not zoning all four corners similarly—or justly.

The default requirement of law is justice in the sense of equality, equity, and fairness. For planning, social utility or efficiency is the starting requirement, or the default position. How to handle consequent injustices is a perennial problem in planning that unfortunately also needs to be addressed. For both law and planning today, justice is measured by standards enacted in properly adopted legislation. If an injustice has occurred, it can typically be remedied by the redistribution of money as damages. In the Bronx Terminal Market case outlined by Susan Fainstein (see her chapter in this volume), the question for planning should be how to achieve social utility while any resultant injustice should be dealt with by law. The question for law would be whether any argument for social utility, approved by acceptable legislative processes, justifies the inequitable result. If the result in the Bronx Terminal Market case was indeed unjust to particular parties, and the court challenges that were launched by the older merchants did not persuade the courts that under prevailing law the result was unjust, then the prevailing law was not implementing principles of justice as it should (and indeed often does not, but that is neither here nor there), and the prevailing law should be changed. The Just City would make justice and equity as much a part of social utility—and a concern of planners—as it is now of law.

Thus, if distributive justice is the central concern, there is no fundamental reason why the matter could not be seen as one of law, and resolved by quite conventionally accepted principles of jurisprudence without regard to any of the broader issues of the common good or capabilities that Fainstein's chapter also raises. It would not be hard to imagine laws that would incorporate the principle of distributive justice in a case such as the Bronx Terminal Market. Donald Hagman's "winners compensate losers" (Hagman and Misczynaski 1978) would be an obvious possibility. By such a measure a just solution to the Bronx Terminal Market case could easily be visualized through any of the alternatives actually proposed. A "winners compensate losers" rule is one that could be implemented without significant rupture in any other part of the governmental administrative or planning process. Presumably we would thus be moved closer to a Just City.

But in a good city, would such a change in the law, such an implementation of the concept of justice, bring us closer to justice formulated in terms of broader goals? It would, indeed, reduce the level of injustice, and since the absence of injustice is characteristic of the Just City, it moves us closer to that characteristic. But it would not deal with the poverty of the older merchants, with the lack of employment opportunities for the neighborhood residents, with the pollution imposed by the traffic overhead, with the lack of shopping alternatives for nearby residents, all of which ought to be addressed, as Fainstein recognizes implicitly in advancing the capabilities approach (see Nussbaum 2000). Something more is needed, but to move from remedying injustice to opening the question of what a city should be involves much deeper issues than those addressed by focusing on the immediate issues of justice.

Fainstein's chapter recognizes the issues and pushes in a direction beyond justice in calling for "identification of institutions and policies that offer broadly appealing benefits" (p. 34). "Broadly appealing benefits" need to be defined by some standard other than justice. The "counter-institutions" that are called for are not merely to avoid injustices, but to provide other benefits. They may be required to deal with: (a) what "appealing benefits" a good city should provide and (b) what the structural sources of injustice are. In particular, counter-institutions should deal with those existing institutions and policies that lead to the "imbalance of power" Fainstein cites as creating the injustices revealed by the Bronx Terminal Market case (see Steil and Connolly in this volume for more on the role of balancing power through counter-institutions). If planning for the Just City, or Justice Planning, is to deal with these issues, it must deal directly and explicitly with the question of power and its sources.

The question of power

Commons Planning, as discussed below, differs from Justice Planning in dealing directly with the issue of power, not simply with remedying the results of its use in the distribution of the benefits of society. A preliminary word on the definition of power is thus necessary, differentiating it from exercise of authority.

A society without power need not be a society without order, just as the presence of power is not an adequate guarantee of order, as the United States military is rediscovering in Iraq. The philosophic issues are complex, but the distinction between power and authority is central. Power is the ability to have others do one's bidding, against their own interest and for the benefit of the holder of power. The power relevant to planning is socially created and results from the social, economic, and political structures of a particular society at a particular point in its history. Authority is likewise the ability to have others do one's bidding, not for the benefit of the holder of authority but for a collective benefit, and is put in place by rules agreed to collectively. Authority relevant to planning is expected to promote the general health, safety and welfare. Its rules are based on reason, perhaps democratically adopted in the Habermasian sense. Rules imposed by the powerful differ from those imposed by those with authority by virtue of their greater power and their own benefit.

The crucial difference is that power is based on social, political, and economic inequality among people, inequality which is socially created and structurally embedded. The holding of power involves a socially created relationship of domination and subservience among individuals and groups. A just action is one not affected by such relationships. The point is important in the practical world of planning because it highlights the conflictual character of injustice. In unjust situations there are not only losers but also winners benefiting from their loss. Thus, measures benefiting the poor will be at some expense to the rich and consensus is not a likely solution to such issues of injustice since win–win situations are rare.

Much of planning theory deals with the solution of problems as if relationships of power did not exist. Or, in its more politically conscious forms, it deals with how to adapt solutions designed under non-power conditions to circumstances where the impact of power is obvious and not challenged.

The introduction of justice as a central criterion for planning or public action must inevitably go beyond individual instances of injustice to challenge the legitimacy of the use of power itself. It thus raises questions that go well beyond the scope of the mainstream of planning theory discussions today, and opens much larger questions for confrontation by planners, public officials and citizens in political practice.

The need for structural change

The factors that influence the existence and distribution of power in society as a whole are ultimately structural. Thus, if relations of power are to be addressed, structural change must be addressed. The old jingle quoted at the opening of this essay highlights the importance of this point with concern to the relationship between justice and a Just City. The jingle comes from the early days of the enclosure movement, often taken as the mark of the transition from feudalism to capitalism and an industrialized market economy. It poses the difference between simple justice and structural inequities sharply. It is worth repeating:

> The law locks up the man or woman
> Who steals a goose from off the common,
> But leaves the greater villain loose
> Who steals the common from the goose.

Bertell Ollman (2006) quotes this jingle and does a riff on the difference in academia between Goose From Off the Commons studies and Commons From Under the Goose studies. I would take determining the ownership of the goose to be a matter of distributive justice and the defense of the common to be the broader matter; I would suggest that the need for Justice Planning studies has to be complemented by attention to Commons Planning studies.[1]

If we change the normative goal of planning from one calling only for distributive justice to one challenging the existence of those relationships of power in the society that give rise to injustice, we have to go beyond the Just City as the city of just distributions within existing relations of power. The Just City discussion should not fudge the issue. On the one hand, the focus of Justice Planning on distributive justice suggests that a Just City is simply a city without injustices, accepting the existing structures, laws, and institutions as given. Fainstein's discussion in her chapter in this volume is clearly uncomfortable with that result, as the earlier quoted references to counter-institutions suggests, and opens the door to going beyond remedying existing injustices as the sole guiding principle. Commons Planning must go through that door and consider alternatives to existing relationships of power, otherwise conceived as the differential support for capabilities or the full meaning of the Lefebvrian "right to the city" (see Dikeç and my Postscript in this volume for more on this discussion). A reexamination of the Bronx Terminal Market case cited by Fainstein can help make concrete the argument for a move from Justice Planning to Commons Planning.

The Bronx Terminal Market case and Justice Planning

The case that Fainstein presents in her chapter dealing with the Bronx Terminal Market is not an atypical one. What are the injustices she sees here? They are essentially the inequitable results of public action changing land uses in the Bronx through lease agreements for city-owned property, city financing and zoning, which redound to the benefit of a large, well-connected and well-financed developer and to the detriment of small, long-standing, much poorer, largely immigrant merchants now using the land and their employees. Subsidies are afforded the developer, not the older merchants; city land is used for the benefit of the developer, and no relocation sites are offered the merchants. Was undue influence used to obtain governmental support for the project? Were the older merchants deprived of due process? Were the rich benefited at the expense of the poor? Should the benefits of government action have been differently distributed? The question posed is, in other words, "Were these outcomes unjust"?

The principles by which the essay would resolve these results call on the concepts of justice or equity, two terms which Fainstein uses interchangeably. The definitions of justice discussed start with that of John Rawls, who focuses on fairness and equality in the distribution of primary goods. Fainstein then runs through other potential measures, such as diversity, culminating in a discussion of the capabilities approach, which at one point, however, also ends up with an "alternative that benefits the least well off," the more or less Rawlsian definition with which the discussion began. The solutions to the Bronx Terminal Market case under these principles of justice deal with the benefits (and presumably costs) of public action, which here focus on the use of land and the allocation of subsidies. The ideal solution in the Bronx Terminal Market case could be the integration of the older merchants' market into the new proposed mall, presumably at some cost to either the developer or the government. Perhaps a second-best solution would be allocation of land elsewhere for the older merchants and the provision of relocation support.

In any case, the question is one of the distribution of outcomes, and the solution is one that takes the *status quo* ante as acceptable, since presumably one acceptable immediate solution, if not the ideal one, would be simply killing the project and leaving the situation as it was before the developer appeared on the scene. Process questions are raised, but apparently independent of outcomes, since better processes do not necessarily lead to just outcomes. Can this analysis lead to a more just distribution of the benefits of public actions or, that is, to better outcomes? Yes, and it is certainly worth pressing. Is it really illuminated by the discussion of the broader issues, will it open up broader questions of social justice, raise questions about the ultimate goals of public action, lead towards a vision of the Just City or move to its implementation? Not unless pushed much further. The Just City goal indeed opens the door to such a broader approach, but does not go through the door. For that, we need to move from Justice Planning to Commons Planning.

COMMONS PLANNING

The Bronx Terminal Market case and Commons Planning

The broader questions that the Bronx Terminal Market case raises are: who should ultimately benefit from the Market? What processes should be used to make decisions about it? And what institutions are needed to implement them? If, as the jingle suggests, the answers to these questions lead to an elaboration of the role of the commons, then further questions are raised: Who and for what purpose are decisions that affect the common welfare made? What are the fundamental problems at the Bronx Terminal Market? Who should have the right to use land? What role should the market play? How are private and public interests to be reconciled? What procedure for making land use decisions would best ensure the most desirable result? These are Commons Planning issues.

More specifically, the questions that could be asked might include: Why are not the workers at the Market whose jobs are threatened guaranteed a decent living and a decent job so they do not have to worry about the Market closing? How should the economic insecurity of unskilled workers, the social limitations that left them unskilled, the plight of the 83 merchants who already left the Market, be dealt with? Why do the remaining 17 merchants have to fear for their livelihoods? Does the formal planning process permit serious and informed public participation in answering these questions? What should the role of money in the democratic process be? Is the city's motivation to increase tax revenues valid as the driving force for changing land uses? To what extent does the private ownership of property conflict with arrangements for the use of land for the common good?

The answers to these questions go well beyond what practical planning can achieve in a specific case. They suggest that community-based interests and decision-making processes must be made formally binding on development, subject to general rules democratically decided on, as many proponents of community-based planning urge. They suggest that those rules must provide that most major land use decisions need to be considered in the light of common welfare of the city as whole, with community decisions only being overridden on clear showing of greater common benefits. They suggest that the rights of private property must be seen as endowed with a social purpose, rather than as matters guaranteeing private benefit, as the new Brazilian Law of the Land provides (see Maricato in this volume) and the old Weimar constitution provided. They suggest that economic growth should not be equated with the common good, and that city agencies dealing with economic growth must be limited in their jurisdiction by the broader concerns of other municipal agencies, including those dealing with education, incomes, environment and family welfare, among others, in this particular case.

Addressing structural change

These are structural questions which Commons Planning would raise, although they are not likely to be effectively resolved in a single case such as the Bronx Terminal Market. But they can be raised there. They can be used to broaden the discussion, to begin to challenge the legitimacy accorded to existing practices, to place those in power in the defensive position of justifying their approaches. They can be used to seek out allies, to build coalitions to marshal the political strength to change structural arrangements. In specific cases, depending on the strength of the coalitions assembled, they can in fact not only help to win a particular case but help prevent a repetition, and in the long run produce further changes in the thinking about such problems that will lead to new conceptions of the common good and the importance of its pursuit.

Raising such questions in any concrete case is a tall order. The best of traditional planning, and certainly good Justice Planning, would attempt to be long-range and comprehensive and provide for just distribution of resources; this is called for (if not among the enforceable requirements) in the existing Code of Ethics of the urban planning profession. But rarely is that mandate extended to call for questions such as those Commons Planning raises. The Bronx Terminal Market example hardly hints at them, for Justice Planning only deals with the individual case in the context of existing relations of power. To define the Just City simply as a city in which each individual case is dealt with justly, without questioning the structures and sources of power in which they are embedded, loses the forest for the trees, loses the common for the individual interest. Justice deals with everyone following the rules, a truly Just City must deal with whether the rules are just. The rules must deal with more than justice; they must contribute to the broader goals societies should achieve for all their members.

Seeing the underlying questions

We know enough about how cities function, what institutions and structures underlie their functioning, how and why decisions are made, and how power is exercised and by whom to be able to frame the key questions that need to be asked to go beyond individual cases of injustice and address broader goals. The questions suggested above in the Commons Planning approach to the Bronx Terminal Market are examples, for that particular instance, of how broader issues can be raised. Those questions can be generalized, and while each case needs to be addressed in its own concrete terms, there are common questions likely to underlie most, and we know enough to be able to suggest some of them. In what follows, since the use of space is a key component of most urban planning controversies, the questions are phrased in terms of issues concerning land use.

Crucially, the questions might start by asking: What is the purpose of public action in a particular case? Is it simply to find the highest and best use for a piece of land, or to raise tax revenues for the city, or to promote one business activity or another? Or is it to serve the common good, to improve the lives of individuals that are now or might potentially be affected by the public action? By raising the question of what a city should be and for whom as the starting point of the debate—and the Just City can be a useful concept here—it sets an entirely different framework for the discussion. It changes the terms of the discourse, as the Fainstein chapter suggests is needed. A given issue becomes not one of simple land use, but of the goals of public action. Justice is then a minimum requirement, the contribution to the general welfare the maximum. And the inquiry must ask why these criteria are or are not met.

Phrasing land use questions more broadly in terms of the common good will inevitably lead to examining the processes by which the uses of land are determined in the city. This would thus raise the further issue of the appropriate role of the market in allocating land, the extent of democracy in political decision-making on land use, including the role of developers in the process, and how priorities are established in the expenditure of public funds used for subsidies. Commons Planning would look at all groups having an actual or potential interest in the outcome (potential here is very important, for the standard "stakeholders" definition of those needing to be involved rarely considers them), and would ask what arrangement for the use of this particular site would best serve the interests of those groups, and how conflicting interests might best be resolved. Should the decision be framed in land use terms to begin with, as so many are, rather than in terms of the common good? What are the consequences of the dependence on private developers in a market society, of the private ownership of land to begin with? What does it take to have a knowledgeable citizenry in the first place? Why are those benefiting from market decisions favored over those who might benefit from socially oriented public decisions, both in this case and generally? These questions are all inherent to most land use issues.

Above all, Commons Planning would then ask: What exercise of power is involved in creating the situation to be dealt with, and who is exercising that power? At whose expense is it being exercised, and what potential strength can be mobilized against that exercise? Should not a solution be sought which will be independent of the exercise of power? These are questions that will resonate with the individuals and groups most affected, regardless of their sophistication or philosophic inclinations, and they can be powerful questions in a public debate over what should be done.

Such a Commons-oriented approach is not inconsistent with the approach of Justice Planning. Indeed, pursuing Justice Planning to its logical conclusions leads centrally to questions of power and thus to Commons Planning. The questions specifically raised with regard to the Bronx Terminal Market are a case in point. Justice Planning, in that case, should lead to an examination of the causes of the problems to be addressed but, as proposed, puts that examination aside or uses it only in the service of achieving justice in that particular case (dealing with the theft of the goose). Commons Planning would focus on the larger question of which that is a part (the protection of the commons) and examine the relations of power that have produced the problem, examining the structures that power has created to sanctify, in this case and generally, the role of the market and the way in which the holders of power dominate the state and discourage socially oriented public actions that interfere with the market. The two approaches are not inconsistent, but they differ significantly.

Issues of scale

Once the door is opened to such consideration of broader structural issues underlying individual cases of distributive injustice, other issues become clear. One is the question of scale, the level at which problems need to be tackled. Fainstein argues, properly, that "justice is not achievable at the urban level without support from other levels" (p. 21). If the focus is only on what can be done today by a city government then that is a fair statement. But if the focus is on achieving a good city, a Just City, one supporting the full development of human capabilities, then the focus needs to be broader. The arena in which an alternative will be achieved cannot simply be the city. Both theory and practice teach that what happens in any city is highly dependent on what happens in its region, its nation and the world. Beyond this, not only are solutions dependent on the support of other levels, but also the causes that need to be addressed are to be found at other levels. The structural arrangements creating problems within cities are determined at higher levels—national and increasingly global. Efforts for change often focus on city policies because this is today an arena in which the social forces for change have their greatest strength, and it is the arena in which urban planners and those engaged in the everyday struggles of life find themselves embroiled; it is certainly the arena in which urban planners customarily operate. But it takes no controversial or sophisticated analysis to realize that questions of subsidy and redistribution, of the administration of justice, of economic regulation, of environmental controls, of war and peace, cannot be handled simply at the city level. More than justice is needed, and it is needed well beyond the city limits.

It may seem quixotic to raise such a question when an immediate decision on a pressing concrete case is required, particularly under a national administration apparently impervious to calls for social justice. But the lesson that national policy needs to be changed if the common good is to be served is a lesson that needs to be reiterated at every opportunity. Even in the medium range, ready, effective proposals may have a chance, given political changes, and may unite other forces, groups and institutions around a common political program with benefits in organizational strength and power, if not in immediate results. Fainstein points out that significant change requires social mobilization with effective social movements behind it. Each case of individual injustice can be used to solidify the recognition of how broad and multifaceted the problems are and how many people need to come together towards goals that are shown to be common to all.

CONCLUSION: FROM JUSTICE PLANNING TO COMMONS PLANNING

To return to the opening jingle: confronted with the theft of a goose, proposals must show the link between ownership of the goose and ownership of the common. Distributive justice deals with the goose. To protect the commons, broader questions need to be faced—but those are not very practical questions. It would be irresponsible not to propose immediate action that can be done within existing circumstances and irresponsible to talk of the commons and let the goose go. Those of us who teach planning need to equip our students to do the best they can within the context of real possibilities, and we should not tell them to tell others that something is realistically possible when it is not. But that does not equate with being quiet about the threat to the commons. It means rather seeking approaches that raise the larger issues as well. Certainly the undemocratic nature of the planning process can be raised, as well as the influence of money and political power in making decisions. Alternatives can be proposed. If subsidies are needed, where else subsidies go and what the net distribution looks like can be exposed. Who benefits and who suffers is always an important part of planning analysis and can be eye-opening. The description of the Bronx Terminal Market case in fact illuminated these; they include, but transcend, the simple call for justice.

But that is a rare approach. The problem is that urban planning is essentially goose-oriented planning or, at best, it is Justice Planning. It deals with immediate issues, immediate actions within the existing distribution of power and legal regulatory schemes. It is this pattern that I think Fainstein is trying to break in order to lead planners both to deal with the immediate and with the longer-term normative issues. In practice, that means recognizing the limitations of planning, but not surrendering to them.

Searching for the Just City demands addressing the problems of the commons as a whole. It is fitting and proper that we address the problems of the goose and its theft, and do what we can to see to it that both are dealt with justly. But to achieve a better city in a better society, we need something more than justice for individual parties. We need to deal with the ownership, control and use of the commons. Just access is part, but only part, of that challenge. Planners must keep that in mind, and help demonstrate what it means. Implementing the Commons approach is a task not just for planners, but for citizens, including all residents and users of the commons. Planners should see both the broad potentials and the real limitations on their roles in that process.

NOTE

1 Commons studies need not be concerned with the so-called tragedy of the commons. For a devastating rebuttal of the thesis promulgated by Garrett Hardin, see Ian Angus (2008).

REFERENCES

Angus, I. (2008) "The Myth of the Tragedy of the Commons," Toronto: Socialist Project. Online. Available HTTP: http://www.socialistproject.ca/bullet/bullet133.html (accessed August 25, 2008).

Hagman, D.G. and Misczynski, D.J. (1978) *Windfalls for Wipeouts: Land Value Capture and Compensation*, Washington, D.C.: American Planning Association Planners Press.

Nussbaum, M.C. (2000) *Women and Human Development: The Capabilities Approach*, Cambridge: Cambridge University Press.

Ollman, B. (2006) "Why Dialectics? Why Now?," *Synthesis/Regeneration*, 40 (Summer).

6 As "just" as it gets? The European City in the "Just City" discourse

Johannes Novy and Margit Mayer

INTRODUCTION

Recent decades have witnessed profound changes in the way scholars and activists on the left have approached the issue of social justice and the city. Claims for egalitarian redistribution and social reform, once paramount cornerstones of political contestation from the left, have not entirely faded away. But the defeat of "really existing socialism" and the demise of Keynesian social democracy in the face of neoliberal globalization as well as a diffusion of struggles for the "recognition of difference" (Fraser 1995, 2000) have raised doubts about these old and trusted goals on "the left." Though recognizing gender, sexuality, race, ethnicity, and even lifestyle as bases of difference, scholars writing from a political-economic perspective have found it increasingly difficult to frame justice in terms of diversity as well as equality, and for this reason have frequently limited themselves to offering critique without formulating specific criteria for what exactly defines social justice or a Just City (see Fainstein 2001: 885).

In their efforts to bring a richer dimension to movements for urban justice, scholars rallying behind the banners of hitherto denied differences and identities have meanwhile—often unintentionally—ended up playing into the hands of the neoliberal assault on egalitarianism, thus contributing to the marginalization of struggles against economic inequalities that had represented the central aim of previous generations of scholars on the left. Still others have in fact abandoned the quest for a socially Just City altogether, often by appropriating the language and policy prescriptions of the "Third Way" or by subscribing to New Urbanism's design-oriented "solutions" to cities' problems. In addition, the failures of utopian urbanism in the capitalist West during the twentieth century have also affected scholars' thinking about social justice in the city. Discredited by the failures of past experiments of social and urban reform (Pinder 2005), the utopian spirit that had carried earlier generations of urbanists to sweeping calls for reform have been replaced by pragmatism and caution, and those who continued to call for radical change were mostly belittled or ignored. Particularly progressive urban planning suffered a crisis of confidence (Fainstein 2001). Today, however, against the backdrop of roll-out neoliberalism (Peck and Tickell 2002) and intensifying inequalities within and between cities, calls for alternative, progressive visions of the city are back on the urban agenda (see Mitchell 2003; Lees 2004; Pinder 2005). Susan Fainstein's writings on the Just City are a case in point. Revealing a growing unease about the sidelining of demands for justice in urban planning and development discourses, Fainstein's work has triggered a renewed debate on how to think and act on contemporary urban problems and how justice on the urban level might be achieved (see Connolly and Steil this volume).

THE JUST CITY MODEL: A PATH TOWARDS GREATER SOCIAL JUSTICE?

Fainstein's writings on the Just City are inspired by her long and enduring concern about—and discontent with—the American city and the development patterns that shape it. In fact, even though her broader discussion of justice in relation to planning deals with theory and practice beyond American shores, her exploration of the shortcomings of contemporary urban development and discussion of what better cities might look like is clearly derived from—and chiefly concerned with—developments that are intrinsically tied to American cities. Embedded in a broader discussion of the role of justice in planning, Fainstein particularly takes issue with urban economic development policy in American cities for prioritizing growth, which has historically inhibited moves toward progressive ends and continues to stand in the way of greater social and environmental justice.

At the same time, Fainstein is also disturbed by the way progressive urban scholarship has responded to injustices inherent in urban development processes: "many scholars in the political-economy tradition offered critique without formulating specific criteria of what was desirable" (2001: 885); while those scholars that went beyond critique either viewed themselves as advocates for particular social groups or, due to their preoccupation with issues of participation, sometimes neglected the fact that better planning processes do not necessarily lead to better planning outcomes. Utopian idealism, on the other hand, is not Fainstein's cup of tea either. For her, planning theory is about "the analysis of the possibility for attaining a better quality of human life within the context of a global capitalist political economy" (2000: 470) and not about transforming the world we live in altogether. "While utopian ideals provide goals toward which to aspire and inspiration by which to mobilize a constituency, they do not offer a strategy for transition within given historical circumstances," Fainstein argues (this volume: 28). She posits that planners, policy makers, and activists committed to progressive urban change instead require a vision of what is desirable and feasible within the circumstances in which present-day cities are embedded.

Yet what is "just" and what is "possible" under the current conditions of advanced globalized capitalism? In several of her writings, Fainstein has suggested that Amsterdam represents the "best available model of a relatively egalitarian, diverse, democratic city, with a strong commitment to environmental preservation," i.e., a city that has managed to find a balance between the "values"—democracy, equity, diversity, growth, and sustainability—she deems necessary for social justice to be achieved (Fainstein 1999: 259; 2000; this volume). This depiction resonates with a more general image of Amsterdam as an "ideal city" which is widespread among planners and urban scholars in Europe and beyond.[1]

In this chapter we take issue with the popular portrayal of Amsterdam as a role model that could guide our debates over social justice and progressive urban agendas. We are thereby less concerned with Amsterdam as such than the type of city with which Amsterdam is associated: the model of the European City. Recent years have seen a proliferation of scholarly output that has advertised the "traditional" European City, often ideal-typically opposed to the "American City", as representing an urban model with the normative force that can guide political action in the present context of globalization and European integration (Hoffmann-Axthelm 1993; Bagnasco and LeGalès 2000; Häußermann 2001, 2005; Hassenpflug 2002; Kaelble 2001; LeGalès 2002; Siebel 2004; Häußermann and Haila 2005). Great hopes are attached to the European ideal type for cities, particularly with regard to overcoming social exclusion and segregation but also with regard to supposed positive relations between ecology and density, as well as competitiveness and—though vaguely defined—social and spatial coherence. The European Union meanwhile has also undertaken efforts to revive certain qualities associated with "traditional" cities in early modern European societies and promulgates a normative vision of Europe's urbanity as a remedy to the negative effects of globalization and alleged guarantor for cities' future well-being (see, e.g., Frank 2005).

Our concern is that references to Amsterdam as a model of urban development beyond the European context lend credibility to such arguments even if many of them tend to be characterized by an idealization of European cities' past and an insufficient consideration of their present-day problems and the roots of those problems. Further, using European cities as a point of reference for alternative and progressive visions of urban development distracts from and, worse, naturalizes the inequalities and injustices which mark them—albeit in different forms. We thus lose sight of the larger picture that needs to be addressed if conditions are to be structurally improved. It also tends to undermine the struggles of activists in European cities who are fighting against the manifold forms of injustice, repression, and exclusion that shape urban space across Europe. This chapter details and substantiates this argument in order to explain why references to European cities as well as the normatively charged model of *the* European City in the present-day context run the risk of serving an agenda that constitutes a barrier rather than a facilitator of progressive urban change. By pointing to the recent blossoming of local governance experiments in Latin America, this chapter concludes with an attempt to push the important discussion that Fainstein's work has sparked beyond its current focus on (and problematic dualism between) North American and European cities. Given how its current formulation is essentially rooted in European and North American assumptions and ideas, the search for models of transformative urban action should be extended into other parts of the world to inform and broaden the scope for progressive activism in American and other "first world" cities.

THE EUROPEAN CITY AS AN IDEAL TYPE AND IDEALIZED CITY MODEL

The current appreciation for Amsterdam comes at a time of renewed, broad-based interest in Europe's cities and urban traditions. This interest is exemplified by the success of anti-modern approaches toward urban development in which reactionary designs based on the traditional European city model by architects like Leon Krier are described as new and innovative. And it is illustrated by a proliferation of scholarly writings premised upon the Weberian notion of the European City that attempt to revive the heritage of urban Europe as a guiding framework for political action in the current context of globalization and European integration.

The revived interest in European cities' historic spatial qualities and the proliferation of visual representations of middle- and upper-class nostalgia as a reaction to the deadening trends of homogenization and segregation of uses characteristic of modernist planning is only of peripheral concern to us here. Much of what is built in European cities nowadays makes use of the characteristics—both aesthetic and structural—of "traditional" European cities as a point of reference. Urban development is called upon to respect—or reconstruct—cities' historical urban morphology, including a variety of building types and uses, among which so-called "public" (but in practice often private) spaces are privileged, and to emphasize architectural design that reflects selected elements of individual cities' building traditions and identity. But as Fainstein also points out, calls for different urban forms do not suffice to move towards more socially just, politically emancipatory and ecologically sound cities. At best, they ameliorate urban conditions. At worst they amount to caricatures of European cities' past that camouflage their present-day problems and serve the interests of real-estate developers, urban boosters, and other local elites.

But it is, as suggested above, not only in the "Back to the Future historiography" (Lehrer 2000: 102) of conservative architects and planners that the "European City" has made a comeback in recent decades. A new future for the model of the European City is also proclaimed by urban sociologists, who appeal to the unique traditions of European cities and to the historic and geographic peculiarities of the European urban system (see, e.g., Bagnasco and LeGalès 2000; Häußermann 2001, 2005; LeGalès 2002; Häußermann and Haila 2005; for a discussion see also Kazepov 2005). Bagnasco and LeGalès

(2000) were among the first to argue that greater attention should be paid to the heritage of European cities and the continent's urban system. Invoking Max Weber's notion of the European City and comparing European cities to their North American counterparts, Bagnasco and LeGalès claimed that, despite their obvious diversity, cities in Europe represent a distinct city type and that many of the morphological, legal, social, and cultural characteristics that traditionally defined them were experiencing a revival in the present context of globalization and European integration. While others diagnosed a "fading charm" and the marginalization of European cities as a result of globalization processes (Castells 1996: 431), Bagnasco and LeGalès saw European cities as more resilient to the effects of global change than previous accounts suggested and contended that particularly the diminishing significance of nation-states (as a consequence of globalization and European integration) was actually playing out in their favor.

Echoing Weber's classic inquiry into medieval cities, Bagnasco and LeGalès highlighted cities' historic role as self-managed "local societies" that provided the cultural, social, and legal framework for the rise of the *bourgeoisie* and modern capitalism. Typically characterized by a self-governing urban citizenry (*Bürgertum*) and a high degree of autonomy, cities had been powerful social and political forces before the emergence of territorial states diminished their significance. Now, as a result of the "loosening grip" of nation states, cities' "room for maneuvre" was growing again and offered them opportunities to "create their own identities" and regain their role as key determinants of European societies (Bagnasco and LeGalès 2000: 7).

While they may have exaggerated the extent to which such claims have been made, Bagnasco and LeGalès provide an important corrective to accounts that overstate the homogenization pressures associated with *supra*-local phenomena like globalization. Moreover their work represents a welcome reminder that processes like globalization manifest themselves differently across space and must therefore be explored in light of the institutional and regulatory context they impinge upon. As an analytical framework, however, the ideal-typical notion of the European City remains vague. While the notion of cities as actors is problematic in itself (see Marcuse 2005), Bagnasco and LeGalès also devote relatively little attention to the factors that constrain the degrees of autonomy which present-day cities in Europe have, or rather the autonomy of the collective and individual actors that shape them. Bagnasco and LeGalès are aware of these constraints, to be sure, but they focus on the opportunities recent restructuring processes have provided for cities' capacities to act autonomously. The same holds true for Bagnasco and LeGalès' discussion of the more general dynamics that characterize contemporary urban Europe. The challenges contemporary European cities face—not least growing social inequality, poverty and polarization—are not neglected, yet the emphasis lies on what the authors perceive as European cities' past and present strengths. Emphasis is thereby placed not only on European cities' increased autonomy, plural self-governance, and harmonious coexistence and sense of solidarity among their citizenry, but also on their physical compactness and mixed-use character, civic life, social cohesion, as well as their political culture and local identity.

Most scholarly writings that followed Bagnasco and LeGalès' example and embraced the (neo-)Weberian thesis of the European City are also predominantly concerned with European cities' positive characteristics (Kaelble 2001; Hassenpflug 2002; Häußermann and Haila 2005). Häußermann and Haila (2005: 61) are quite frank about this bias. For them "the neo-Weberian framework is a conceptual framework as well as a normative one" as it "disclose(s) the good qualities of European cities and . . . emphasize(s) the political role of cities, together with the political role and responsibilities of citizens." Here it becomes clear that the revived interest in the "Europeanness" of these cities is not only about creating a heuristic for social scientific analysis. Rather, most scholars who invoke Weber's notion of the European City conceive of its "traditional" form as an historic achievement and possible blueprint to guide future urban development. And they at least implicitly embrace the ideals associated with the rise of the *bourgeoisie* to which the notion of the European City is intrinsically tied. Once emancipatory forces against the powers of feudalism and clericalism, these ideals have taken on a different meaning in today's political context (Frantz 2007: 17). Now they are frequently used to promote free markets, individualism and individual responsibility, and/or "civic" responsibility as opposed to state solutions as responses to societal problems.[2] As a result, in today's political context they do not necessarily represent a facilitator of progressive social change.

Actually existing European cities and contemporary urban restructuring

In the present historical conjuncture, characterized by neoliberal restructuring and a reconfiguration of the relationship between (nation) state and civil society, the division of labor between municipal and higher levels of government undeniably has changed. As more and more responsibilities that used to be centrally organized have been devolved to local administrations, cities have acquired much stronger roles in the political, economic, social and cultural arenas, as Bagnasco and LeGalès suggest. They have done so, however, under circumstances that hardly allow them to advance principles which go against the grain of the prevailing ideologies and practices of what Brenner and Theodore (2002) call "actually existing neoliberalism"[3] and other variants of new-right conservatism. The current erosion of civil rights and liberties through steadily expanding incursions of state power and surveillance, the growing exclusion of the long-term unemployed, as well as the increasing discrimination and criminalization of migrants and groups with migrant background represent realities of discrimination and injustice throughout urban Europe against which local governing arrangements, even when fundamentally opposed, have little leverage. Local government leaders have had little choice but to abandon welfarist social policies in favor of an entrepreneurial, competitiveness-oriented approach, as globalization, economic restructuring and neoliberal policies implemented by national and supranational institutions—such as the European Union—have changed the institutional-territorial matrix of urban governance. These higher-level policies have shifted public resources from universal and redistributive programs to targeted territorial projects geared at supporting supply-side intervention, central-city makeovers, place-marketing efforts, and similar activities (see Hall and Hubbard 1998). Hence policies and programs addressing severely "disadvantaged" people and areas are no longer redistributive, but tend rather to simply combine market-driven processes with civic involvement and "capacity-building" efforts even while "liberalized" housing markets destroy the right to housing for more and more groups and public spaces have become more highly policed, robbing, for example, homeless people of basic rights, such as the right to beg or the right to lie on sidewalks (see Mayer 1994, 2007). Competitive logics and privatized management have been extended into spheres that until only recently were relatively socialized, involving not only the elimination of public monopolies for the provision of standardized municipal services (e.g., utilities, sanitation, public safety, mass transit) but also the creation of privatized, customized and networked urban infrastructures through competitive contracting (see Rügemer 2006, 2008).

Urban governance in European cities, in short, has become shaped by an emphasis on economic growth and efficiency, entrepreneurialism, competitiveness, deregulation, and privatization, as well as a shift away from welfare towards "workfarist" social policies (see Eick *et al.* 2004). Social problems and conflicts have sharply increased as wealth and opportunities have become more unevenly distributed in Europe and new concentrations of poverty and disadvantage have emerged. But these problems are rarely tackled at their roots, not least because of the increasingly dominant neoliberal subjectivity (Leitner *et al.* 2006).

The new generation of frequently European Union (EU)-funded neighborhood management and community regeneration programs that came into existence in the course of the 1990s to address the rising deprivation of urban neighborhoods are a case in point (Andersen and van Kempen 2001; Mayer 2006). Most of them tend to define the problem not as one of rising inequalities and aggravating poverty, but of lack of "inclusion." Hence, low wages or poverty need not be addressed; in fact they do not pose a problem as long as people are integrated into social networks. Instead, "exclusion"—defined as the detachment of particular groups from normal social life—becomes the focus. "Exclusion" is best addressed by "normalizing" those groups and individuals by inserting them into the labor market and changing their behaviors (Gough and Eisenschitz 2006; Mayer 2006). Insertion of these groups into waged employment or micro-enterprise is seen both as the means and the key indicator of inclusion into a society seen as basically sound.

Neighborhoods where they concentrate are seen as compounding and fostering their "exclusion." Hence, comprehensive programs must target these neighborhoods, encourage local partnerships and activate a variety of local stakeholders. Funding for these programs is minimal in comparison with the subsidies and tax abatements for investors and developers promising to push the growth of booming areas. And the institutional landscape and the discourse set up by the new anti-exclusion programs not only distract from the structural causes of the growing inequalities but also deter people from more effective organizing and mobilizing, which actually might reverse the decline and lead to real improvements. These responses to exclusion, in other words, neither seek to decrease the number of the poor nor impact in any way the production of poverty. They problematize neither the "inclusion" into a basically unjust society nor the unjust fashion in which people are "included" into a social arrangement.

AMSTERDAM—ALMOST ALRIGHT?

Of course differences between cities exist. Amsterdam, for instance, is widely seen as a city that resisted neoliberal pressures. Comparative studies tend to place it near the low end of segregation for European cities, social contrasts appear less pronounced, and opportunities for social mobility are relatively large (see Blok *et al.* 2001; Musterd and Deurloo 2002; Musterd and Salet 2003). Social attitudes and government policy concerning immigration and diversity have long been considered progressive, and municipal government has been praised for involving residents and community organizations in policy-making processes as well as responding with tolerance and leniency to squatters and other manifestations of radical opposition (Uitermark 2004). At the same time, however, developments in Amsterdam do not take place in a fortress shielded from the threats of global urban restructuring and its consequences (see BAVO 2007). Barely noticed in the scholarly realm (Uitermark 2008), much of what happens in Amsterdam today conforms to developments in other parts of Europe as Keynesian and collectivist strategies have taken a back seat to neoliberal market ideologies and forces which gained prominence in Amsterdam in the course of the 1990s alongside the parallel erosion of the Dutch welfare state (Nijman 1999; BAVO 2007).

Social housing has been increasingly demolished or converted to market-rate housing (Uitermark 2004; Oudenampsen 2006, 2007, 2008); cultural production has been redirected towards the promotion of tourism, leisure, and real-estate development; welfare measures relating to education and social services have been decreasing; programs and projects geared at achieving economic success in competition with other cities for investments, innovations, and "creative classes," such as the Zuidas project, a new business district south of the city, have come to dominate the public policy agenda; inner-city neighborhoods are gentrified and differences in living conditions between groups and neighborhoods are growing (Kloosterman 1996; Oudenampsen 2007, 2008). Furthermore, Amsterdam's much praised tradition of tolerance has been challenged as opposition to cultural pluralism has increased (e.g., van Swaaningen 2007: 241). For instance, the assassinations of Pim Fortuyn in 2002 and Theo van Gogh in 2004 contributed to a hardening of public attitudes and a reconsideration of "multiculturalist" policies by the government (Uitermark *et al.* 2005). Finally, in sharp contrast to Amsterdam's tolerant image, local research has detected a move towards "urban revanchism" as well as an emerging "culture of control" in recent years as the city seeks to reclaim its image and attract business investments and residents with higher incomes (Oudenampsen 2007, 2008; van Swaaningen 2007: 238).[4] Titled "I amsterdamned—Free city with lots of opportunities," a protest campaign which was launched in response to the city's present marketing campaign "I Amsterdam" summarizes Amsterdam's current political climate as follows (see Figure 6.1):

> Job Cohen, mayor of Amsterdam, is proud to welcome all of you investors in our liberal and tolerant town. In order to make your stay here as pleasant as possible every effort has been made to put a stop to the lifestyles of people that are unprofitable to us. Countless artists, squatters, backpackers, unemployed, social organisations, homeless, voluntary organisations, council tenants, street artists, musicians, illegals and other riff raff have already left this town. Happily waved goodbye by the mayor: Bye Amsterdam!

This is not to suggest that Amsterdam has surrendered all of the qualities for which it has been praised. Utopian Amsterdam, in other words, has not turned dystopian overnight. But the city's recent trajectory suggests that: (a) the Amsterdam of yesteryear never was an embodiment of utopia in the first place, since present conflicts and problems did not emerge overnight; and (b) that Amsterdam is subject to the same political-economic and social restructuring processes as other cities in the advanced capitalist world and deals with them in much the same way. Differences between Amsterdam and other cities are thus of degree and not of kind. With regard to the daily lives of Amsterdam's residents, these differences surely matter as the city's development continues to be mediated through its social-democratic tradition and the still evident achievements of the social movements of the 1970s and 1980s, as well as by what remains of the Dutch welfare state, and by the ongoing struggles of local activists who claim a "right to the city" for the disadvantaged, marginalized, and oppressed. But particularly since Amsterdam has long been hailed as a center of progressive urban policies and the epitome of a tolerant, egalitarian, and cosmopolitan city, its realignment in the course of the last decades nonetheless provides a stark example of the pervasiveness of European cities' contemporary remaking (see Oudenampsen 2006). Instead of providing a possible blueprint for progressive planners,

Figure 6.1 I amsterdamned—free city with lots of opportunities. Photo by *Ravage*, Dutch Action Paper

present-day Amsterdam thus suggests that urban Europe under current conditions might not be the most useful reference point for informing and broadening the scope of progressive activism in U.S. cities. Instead, the current development of European cities and particularly the discourses surrounding the notion of the European City that surround it have to be seen as deeply

Figure 6.2 Resistance to eviction by squatters in Leidsestraat Amsterdam, October 2006. Photo by Karen Eliot

ambivalent. The latter may be well intentioned, especially when used to critique the socially fragmented, environmentally wasteful, and culturally homogenizing features of American-style developments, but references to the historic model of the European City mask the fact that cities in Europe throughout their feudal and capitalist history have been places of both integration and discrimination. Many qualities associated with late medieval and early modern European cities primarily benefited the (ascending) *bourgeoisie* and presupposed the existence of a set of specific—not necessarily enviable—cultural, social, economic, and political conditions. Many qualities of twentieth-century European cities, such as the achievement of a socioeconomically mixed urban population, and the provision of a modicum of social security for needy strata of the population, meanwhile have less to do with cities than with sophisticated national welfare states and their legally sanctioned social policies. With the dismantling of these policies and the wider political-economic framework within which they used to operate, it is difficult to foresee a future for European cities as exemplars of progressive urban change.

CHALLENGING NEOLIBERALISM—LATIN AMERICA'S "NEW" LEFT

Different ways to approach the struggle for greater urban social justice exist. Susan Fainstein has organized much of her inquiry around Amsterdam as providing an "actual model of social justice" from which important lessons can be drawn for planning and policymaking today. This claim is made despite a clear recognition that many of the qualities that led to Amsterdam's image as a desirable urban model relied on a particular historical and institutional context that has either ceased to exist or is currently being destabilized and disrupted. Underlying her argument is the premise that the "[a]chievement of the just is a circular process, whereby the pre-existence of equity begets sentiments in its favor, democratic habits produce popular participation, and diversity increases tolerance" (1999: 266). Somewhat in contradiction to this view, we find that some of the most interesting sources of inspiration for current struggles towards different and more just processes of urbanization have emerged in parts of the world where problems of injustice and inequaity are particularly pronounced and the material and ideological characteristics of neoliberalism have historically been particularly influential (Davis 2006). Just like urban research and theory more generally (Roy 2005), the notion of the Just City has been characterized by a focus on distinctively Euro-American assumptions and ideas. Greater attention should be paid not only to the urban transformations and experiences in the developing world, as more and more urban researchers are acknowledging, but particularly to recent developments in Latin America, as they might provide beneficial examples for the struggle towards greater urban justice in North America and beyond.

A "New" Left in the form of progressive and often radical urban movements as well as local, regional, and national governments has surged in Latin America in recent decades and holds interesting lessons for those of us in the Northern hemisphere committed to progressive urban change (Angotti 2006; Kennedy and Tilly 2008). While local elites in Europe usually base their actions either on the assumption that people will be best off with a democratic, capitalist order that prioritizes market features and minimizes public sector involvement, or on the clichéd claim that "there is no alternative" (the so-called TINA syndrome to the neoliberal *status quo*), the Latin American situation has been different. Here, left-led local governments over the last few decades have, with the support of social movements or in response to their pressure (Souza 2006), in several cases resisted—or reversed—the neoliberal approach to governance and planning (see Maricato in this volume). Irrespective of their broad range of variation, these "city-level experiments" in places as diverse as Lima, Pôrto Alegre, Montevideo, La Libertador in Caracas or Mexico City have shared a commitment to changes directed toward greater social and spatial justice within a redistributive framework of opposition to neoliberalism (Chavez and Goldfrank 2004). As well, they have in many cases instituted new forms of participatory local democracy and bottom-up planning to shift power downward, to reverse citizen indifference and to expand ordinary people's political rights.[5] Though it must be recognized that these experiments are the exception rather than the rule and that they take place within a context that leaves much to be desired, such efforts nonetheless deserve closer attention. Not only do they "represent important steps towards challenging the current orthodoxy of neoliberalism and liberal (or minimal) democracy" (Chavez and Goldfrank 2004: 195) but they also transcend left politics' (electoral and otherwise) earlier bias towards centralized and technocratic governance models, and they experiment with more democratic and accountable forms of policy-making that break with dependency and extend more political rights to the population.

More attention to the Latin American "New" Left involves not only a critical evaluation of local governments' actions but also a closer look at the strategies progressive forces (i.e., social movements, unions, indigenous organizations, and NGOs) employ to pressure, enter, and ultimately prevail in electoral politics despite frequently hostile policy environments on the regional, national, and *supra*-national level. That not all experiences of local left-run governments in Latin America lived up to their promise is well documented (e.g., Chavez 2004; Chavez and Goldfrank 2004), but the flaws and failures that have accompanied the recent waves of reform across the Americas do not alter the fact that Latin America's "New" Left has infused new energy into the struggle against urban social injustices. This is an example of alternative, emancipatory, and transformative politics that a Just City project has much to gain from.

CONCLUSION

This chapter challenges the notion that Amsterdam is an exemplar of "practical possibility" that can guide us in our struggle towards a Just City in the United States and beyond. We argue that, despite its past successes, Amsterdam in the current political-economic climate does not present a model that can inform strategies for realizing more just urban futures in the here and now. We are, moreover, concerned that by painting too positive a picture of Amsterdam's past and paying too little attention to its present-day problems, the Just City formulation runs the risk of lending credibility to attempts to employ the notion of the European City and its real and perceived qualities for purposes that run counter to the idea of progressive urban change.

Ultimately, however, our critique also points to a larger and more fundamental difference between Susan Fainstein's model of the Just City and how it could be achieved and our own perspective. Despite her vocal critique of present-day development patterns and policies, Fainstein remains pragmatic with respect to the current political-economic climate in the advanced capitalist world and considers the existing capitalist régime of accumulation, including the problems of global injustice and ecological sustainability that it entails, to a large extent to be a given. While Fainstein thus seeks to correct persisting injustices on the urban level within the context of a global, capitalist political economy, suggesting that the case of Amsterdam might show us the way, we are skeptical of the assumption that urban social justice and a capitalist order can go hand in hand. Instead, we argue that the Just City approach in its current formulation unnecessarily constrains the struggle for urban social justice, sweeping alternative visions and alternative possibilities aside. Therefore it would seem to us that alternative imaginaries and practices that challenge the current hegemonic order on the local level can better inform our joint struggle to achieve truly progressive urban change in the United States and beyond. Such imaginaries and practices are —at least with respect to public policy making—few and far between in Europe, but they have flourished across Latin America in recent years. Progressive and radical local governance experiments in Latin America involve plenty of imperfections and pitfalls that could not be discussed here. But these experiments deserve greater attention, for they illustrate that urban policy agendas that go against the grain of neoliberalism can prevail—even if only temporarily—and bring improvements—however modest—to the lives of urban residents, particularly the poor. Driving many of these experiments is the assumption that truly transformative change is not illusionary or utopian and is worth fighting for. It is this optimism that we miss most in the current formulation of the Just City.

NOTES

1 While Fainstein in her latest formulation of the Just City (this volume) has in fact somewhat de-emphasized the role of Amsterdam as "best model," referring to the city as providing merely a "rough image of a desirable urban model," a recent conference organized by American scholars, to give an example, maintained that Amsterdam "is known as the ideal city for the 21st Century when measured by the standard of social justice, environmentalism, planning and design" (Gilderbloom 2008).

2 The evolving urban policy of the European Union with its focus on competitive cities and city-regions, which at least for a time made aggressive use of the notion of the European City as a cradle of civilization and democracy, is a case in point (see Frank 2005).

3 In recent years, critical geographers have observed that the transformations associated with neoliberal reform do not conform to a simple model of liberalization

and marketization with predictable implications for spatial transformation. Rather, "actually existing neoliberalism" is shaped by "national, regional, and local contexts defined by the legacies of inherited institutional frameworks, policy régimes, regulatory practices, and political struggles" (Brenner and Theodore 2002: 351).

4 This is exemplified by highly contested raids to round up undocumented immigrants; allegations of racial profiling during stop and search operations of the police as well as increased efforts to "clean up" parts of the Red-light District and drive the homeless out of certain parts of the inner city.

5 Arguably the most prominent example of municipal participatory democracy is represented in the efforts in Pôrto Alegre, Brazil to redistribute city resources in favor of the city's vulnerable social groups through participatory budgeting (see Sousa Santos 1998; Novy and Leubolt 2005). Less prominent but equally notable examples include the municipal decentralization efforts of Montevideo's left-wing coalition *Frente Amplio-Encuentro Progresista* (Canel 2001) and the policies that have been introduced by the Bolivarian government to foster the development of a "participative and protagonist democracy" in the slums of Venezuela's capital, Caracas (Bernt *et al.* 2007).

REFERENCES

Andersen, H.T. and van Kempen, R. (eds) (2001) *Governing European Cities: Social Fragmentation, Social Exclusion and Urban Governance*, Aldershot: Ashgate.

Angotti, T. (2006) "Book Review: Cities in Latin America: More Inequality," *Latin American Perspectives*, 33: 165–174.

Bagnasco, A. and LeGalès, P. (2000) "European cities: local societies and collective actors?" in Bagnasco, A. and Le Galés, P. (eds) *Cities in Contemporary Europe*, Cambridge: Cambridge University Press: 1–32.

BAVO (2007) "Democracy and the Neoliberal City: The Dutch Case," in BAVO (ed.) *Urban Politics Now: Re-Imagining Democracy in the Neoliberal City*, Rotterdam: Nai Publishers: 212–234.

Bernt, M., Daniljuk, M., and Holm, A. (2007) Informelle Urbanisierung, Selbstorganisation und Sozialismus des 21. Jahrhunderts. Partizipative Stadtentwicklung in den Barrios von Caracas, *Prokla*, 37(4): 561–577.

Blok, H., Musterd, S., and Ostendorf, W. (2001) *The Spatial Dimensions of Urban Social Exclusion and Integration: The Case of Amsterdam, the Netherlands*, URBEX Series: No. 9.

Brenner, N. and Theodore, N. (2002) "Cities and the geographies of 'Actually Existing Neoliberalism,'" *Antipode*, 3: 349–379.

Canel, E. (2001) "Municipal Decentralization and Participatory Democracy: Building a New Mode of Urban Politics in Montevideo City?" *European Review of Latin American and Caribbean Studies*, 71: 25–46.

Castells, M. (1996) *The Rise of the Network Society*, Malden, MA: Blackwell.

Chavez, D. (2004) *Polis & Demos. The Left in Municipal Governance in Montevideo and Porto Alegre*, Amsterdam: TNI/Shaker Publishing.

Chavez, D. and Goldfrank, B. (2004) (eds) *The Left in the City: Participatory Local Governments in Latin* America, London: Latin American Bureau.

Davis, M. (2006) *Planet of Slums*, New York: Verso.

Deurloo, M.C. and Musterd, S. (2001) "Residential profiles of Surinamese and Moroccans in Amsterdam," *Urban Studies*, 38(3): 467–485.

Eick, V., Grell, B., Mayer, M., and Sambale, J. (2004) *Nonprofits und die Transformation lokaler Beschäftigungspolitik*, Münster: Westfälisches Dampfboot.

Fainstein, S.S. (1999) "Can We Make the Cities We Want?" in Beauregard, R.A. and Body-Gendrot, S. (eds.), *The Urban Moment: Cosmopolitan Essays on the Late-20th-Century City*, Thousand Oaks, CA: Sage Publications: 249–71.

—— (2000) "New directions in planning theory," *Urban Affairs Review*, 35(4): 451–478.

—— (2001) "Competitiveness, cohesion, and governance: their implications for social justice," *International Journal of Urban and Regional Research*, 25(4): 884–888.

—— (2005) "The European Union and the European cities: three phases of the European Urban Policy," in Altrock, U., Günter, S., Huning, S., and Peters, D. (eds) *Spatial Planning and Urban Development in the Ten New EU Member States*, Aldershot: Ashgate: 307–322.

Frantz, M. de (2007) "The city without qualities: political theories of globalization in European cities," *EUI Working Paper SPS*, 2007(04) Badia Fiesolana: European University Institute, Department of Political and Social Sciences.

Fraser, N. (1995) "From redistribution to recognition: dilemmas of justice in a 'post-socialist' age," *New Left Review*, 212: 68–93.

—— (2000) "Rethinking recognition," *New Left Review*, 3: 107–120.

Gilderbloom, J. (2008) Ideal City Conference; October 11–14, 2008, Call for Papers. Online. Available HTTP: http://www.urbanicity.org/IdealCityAlert.htm (accessed May 12, 2008).

Gough, J. and Eisenschitz, A. with McCulloch, A. (2006) *Spaces of Social Exclusion*, London: Routledge.

Hall, T. and Hubbard, P. (1998) (eds) *The Entrepreneurial City: Geographies, Politics, Regime and Representation*, Chichester: John Wiley and Sons.

Hassenpflug, D. (2002) (ed.) *Die Europäische Stadt—Mythos und Wirklichkeit*, Münster: LIT.

Häußermann, H. (2001) Die Europäishe Stadt. Leviathan 29(2): 237–255.

—— (2005) "The end of the European City?" *European Review*, 13(2): 237–249.

Häußermann, H. and Haila, A. (2005) "The European City: a conceptual framework and normative project," in Kazepov, Y. (ed.) *Cities of Europe*, Oxford: Blackwell: 43–64.

Hoffmann-Axthelm, D. (1993) *Die dritte Stadt: Bausteine eines neuen Gründungsvertrages*, Frankfurt am Main: Suhrkamp Verlag.

Kaelble, H. (2001) "Die Besonderheiten der europäischen Stadt im 20. Jahrhundert," *Leviathan* 29(2): 256–274.

Kazepov, Y. (2005) "Cities of Europe: changing contexts, local arrangements, and the challenge to social cohesion," in Kazepov, Y. (ed.) *Cities of Europe*, Oxford: Blackwell: 3–42.

Kennedy, M. and Tilly, C. (2008) "Making sense of Latin America's 'Third Left' ". *New Politics*, 11(4): 11–17.

Kloosterman, R.C. (1996) "Double Dutch: polarisation trends in Amsterdam and Rotterdam after 1980," *Regional Studies*, 30(5): 467–476.

Lees, L. (ed.) (2004) *The Emancipatory City? Paradoxes and Possibilities*, London: Sage.

LeGalès, P. (2002) *European Cities: Social Conflict and Governance*, Oxford: Oxford University Press.

Lehrer, U. (2000) "Zitadelle Innenstadt: Bilderproduktion und Potsdamer Platz," in Scharenberg, A. (ed.), *Berlin: Global City oder Konkursmasse? Eine Zwischenbilanz zehn Jahre nach dem Mauerfall*. Berlin: Karl Dietz Verlag.

Leitner, H., Peck, J., and Sheppard, E.S. (eds) (2006) *Contesting Neoliberalism*, New York: Guilford Press.

Marcuse, P. (2005) "'The city' as perverse metaphor," *Cities* 9(2): 247–254.
Mayer, M. (1994) "Post–Fordist city politics," in Amin, A. (ed.) *Post-Fordism: A Reader*, Oxford: Basil Blackwell: 316–337.
Mayer, M. (2006) "Combattre l'exclusion sociale par *l'empowerment*: le cas de l'Allemagne," *Geographie Economie Societé* 8(1): 37–62.
—— (2007) "Contesting the neoliberalization of urban governance," in Leitner, H., Peck, J., and Sheppard, E. (eds) *Contesting Neoliberalism—The Urban Frontier*, New York: Guilford Press: 90–115.
Mitchell, D. (2003) *The Right to the City: Social Justice and the Fight for Public Space*, New York: Guilford Press.
Musterd, S. and Deurloo, M.C. (2002) "Unstable immigrant concentrations in Amsterdam: spatial segregation and integration of newcomers," *Housing Studies* 17(3): 487–503.
Musterd, S. and Salet, W. (eds) (2003) *Amsterdam Human Capital*. Amsterdam: Amsterdam University Press.
Nijman, J. (1999) "Cultural globalization and the identity of place: the reconstruction of Amsterdam," *ECUMENE*, 6(2): 146–164.
Novy, A. and Leubolt, B. (2005) "Participatory budgeting in Porto Alegre: social innovation and the dialectical relationship of state and civil society," *Urban Studies* 42(11): 2023–2036.
Oudenampsen, M. (2006) "Extreme makeover," *Mute* 2(4): 38–43.
—— (2007) "Amsterdam, the city as a business," in BAVO (ed.), *Urban Politics Now: Re-Imagining Democracy in the Neoliberal City*, Rotterdam: Nai Publishers: 110–127.
—— (2008) "Back to the future of the creative city. an archaeological approach to Amsterdam's creative redevelopment," *Variant*, 31: 16–19.
Peck, J. and Tickell, A. (2002) "Neoliberalizing space," *Antipode*, 34: 380–404.
Pinder, D. (2005) *Visions of the City*, London: Routledge.
Roy, A. (2005) "Urban informality: towards an epistemology of planning," *Journal of the American Planning Association*, 71(2): 147–158
Rügemer, W. (2006) "Finanzinvestoren greifen nach Wohnungen. Vom öffentlichen Eigentum zum Renditeobjekt," *Forum Wissenschaft*, 4: 19–22.
—— (2008) *Privatisierung in Deutschland: Eine Bilanz*, Münster: Westfälisches Dampfboot.
Siebel, W. (ed.) (2004) *Die europäische* Stadt, Frankfurt am Main: Suhrkamp.
Sousa Santos, B. de (1998) "Participatory budgeting in Porto Alegre: toward a redistributive democracy," *Politics and Society*, 26(4): 461–510.
Souza, M.L. De (2006) "Social movements as 'critical urban planning' agents," *City*, 10(3): 27–342.
Uitermark, J. (2004) "Framing urban injustices: the case of the Amsterdam squatter movement," *Space and Polity*, 8(2): 227–244.
—— (2008) Personal email communication with the authors, June 2008.
Uitermark, J., Rossi, U., and van Houtum, H. (2005) "Reinventing multiculturalism: urban citizenship and the negotiation of ethnic diversity in Amsterdam," *International Journal of Urban and Regional Research*, 29(3): 622–640.
Van Swaaningen, R. (2007) "A tale of two cities. The governance of crime and insecurity in Rotterdam and Amsterdam," in Sessar, K., Stangl, W., and van Swaaningen, R. (eds) *Großstadtängste—Anxious Cities*, Untersuchungen zu Unsicherheitsgefühlen und Sicherheitspolitiken in europäischen Kommunen: 237–257.

7 Urban justice and recognition

Affirmation and hostility in Beer Sheva

Oren Yiftachel, Ravit Goldhaber, and Roy Nuriel

> What we ask from you is simple: just observe the law; if you do this, everybody will benefit: you will have well-planned, serviced, and recognized towns, and we'll safeguard the last tracts of vacant land for the Jewish people around the world, and particularly for those who stayed for the time being in the ex-Soviet Union, for a possible day of crisis.
>
> (Ze'ev Boym, Minister for Housing, Beer Sheva, June 14, 2006)

The context of this quote is the unresolved land and planning disputes between the Bedouin Arabs surrounding the city of Beer Sheva and the Israeli state. The minister asks the indigenous Bedouin in no uncertain terms to leave their ancestors' land, where they reside in "unrecognized" (and in the eyes of most Israeli planners "illegal") villages and towns, and relocate into modernized, legal, and well-serviced localities.

Beyond the colonialist disregard of indigenous rights embedded in the minister's vision, he unwittingly exposed a dilemma about recognition—widely accepted as "positive" in discussions about spatial justice. His comments invoked a type of recognition which works against, not for, group rights and social justice. At the same time, he extended privileged recognition to potential Jewish immigrants. This differentiation provides a puzzling aspect to our thinking about urban justice and group rights rarely addressed by planning theorists. Should we, can we, "open up" the Pandora's box of recognition?

This chapter explores the relations between recognition and justice. We analyze the treatment of various immigrant and indigenous groups by state and urban authorities, and highlight the manner in which various types of recognition guide urban policy. Our central argument takes issue with the mainstream view of recognition as a necessarily positive element in the pursuit of urban justice. Instead, we view it as a multifaceted socio-political process, ranging between positive affirmation, marginalizing indifference, and exclusive hostility, with a multitude of possibilities in between these poles. We argue that the "gradients" of recognition are linked to significant changes in the urban fabric. Not only are they clearly associated with socioeconomic (class) stratification but also with phenomena we identify as new "urban colonialism," "creeping urban apartheid," and the formation of "gray" (informal) spaces.

We thus seek to advance the discussion on spatial justice by opening up the rubric of recognition. We maintain that a more sophisticated and critical understanding of this concept is needed, and that recognition, or lack thereof, may enhance or harm social and spatial justice. Recognition should thus be viewed as a continuum, and governing bodies should be aware of the damaging possibilities of marginalizing indifference or exclusive hostility as well as the positive possibilities of affirmative recognition.

Following a theoretical discussion, a conceptual scheme is used to analyze the impact of planning on various groups in the Beer Sheva region. We trace the formulation of differential policies: affirmative recognition is extended to "Russian" immigrants (denoting Russian-speakers hailing from the former USSR); "marginalizing indifference" is prevalent in the policies towards Mizrahim (Jews arriving from the Middle East and their descendants); and hostile recognition is evident *vis-à-vis* most Arabs in the region. The claims of Palestinian refugees are totally absent from the planning discourse, while potential Jewish migrants, as noted in the minister's statement, cast a distant but ever-present shadow over the allocation of space in the region.

This chapter aims to rethink social justice under conditions of variegated recognition. We briefly suggest below the "right to the city" (Lefebvre 1991, 1996; Mitchell 2003) as a possible guiding principle for combining recognition and spatial justice, while avoiding the colonial pitfalls of planning for different types of recognition. This requires politicization and specialization of the abstract concept and critical engagement with mainstream liberal literature on urban justice.

PLANNING, JUSTICE, AND DIFFERENCE

The story planners are often told paints the profession as rising out of the desperate chaos of the exploding industrial city in order to introduce order, public health, urban organization, and livability. The heroes of this history are men of great vision, who combine new forms of urban living with social agendas of equality, modernity, community, and a new moral and professional zeal (for reviews see Cherry 1988; Hall 1988; Friedmann 2002). It was not until the 1970s that the story was seriously challenged by new political economic analyses of planning, which drew attention to the structural links between planning, economic structure, and development capital. This led to the exploitation of planning by capital at the expense of weaker social groups (Castells 1978; Hague 1984). Some studies have focused on the role of planning as a legitimacy mechanism for the uneven manifestation of the capitalist state (Dear 1981). Others have drawn attention to the social geography of planned cities still deeply fragmented by class, forming the foundation for long-term inequalities (Marcuse 1978; Troy 1981; Badcock 1984). A common theme links the state, and hence planning, with a privileged facilitation of capitalist demands, and a purposeful neglect of social needs, in what Marcuse (1978) perceptively termed "the myth of the benevolent state."

It was on the basis of such critical analyses that an "urban" justice literature began to emerge as an attempt to rethink the links between space, development, power, and planning. David Harvey's (1973) *Social Justice and the City* and Manuel Castells' (1978) *The Urban Question* served as inspiring texts for a new generation of urban researchers seeking ways to realize "a just distribution, justly arrived at" (Harvey 1973: 97). Claims advanced by the new "justice literature" caused intense, often bitter, debates among planning and urban scholars, especially between Marxian and rationalist/liberal thinkers. Yet both camps agreed that planning was essentially about the process of distributing material resources.

However, the apparent agreement over the parameters of just planning did not last. During the 1980s and 1990s, new claims for a Just City began to appear, challenging previous accounts and taking the analysis of urban justice to new spheres. In the main, three related and partially overlapping perspectives informed these challenges: identity, feminism, and postmodernism. The new wave gave rise to seminal works such as Iris Marian Young's *Justice and the Politics of Difference* (1991); Leonie Sandercock's *Cosmopolis* (1998), and Jane Jacobs' *Edge of Empire* (1996). These works demonstrated the necessity of accounting for issues of difference and identity in order to both understand the emerging urban order and to reformulate visions of urban and spatial justice.

Other studies during this wave took planning theory outside the liberal West and highlighted the close links between ethno-nationalism, religion, the state, and the making of cities and regions. They explored the critical role of urban policy in shaping not only class but also ethnic, cultural, and racial relations, in which space is a critical axis (Falah 1989; Yiftachel 1991; Thomas 1995; Bollen 1999, 2007). This is particularly so in "ethnocratic" régimes, which work to enhance the position of a dominant ethnic group while actively marginalizing minorities and peripheral ethno-classes (see Kedar 2003; Yiftachel 2007). Other studies have shown the centrality of race to urban structure and segregation and from these to notions of corrective justice and improved terms of collective coexistence (Thomas and Krishnarayan 1993; Sandercock 1995; Massey 2007).

The main consequence of this discussion was the introduction of new categories and entities into the vocabulary and imagination of the Just City concept, most notably "recognition," "diversity," "difference," and "multiculturalism." Urban and planning theorists did not have to travel far in search of inspiring texts, with the works of Taylor (1992), Hall (1991), hooks (1995), and Kymlicka (1995) offering new philosophical and political foundations for rethinking the just multicultural city.

RECOGNITION AND REDISTRIBUTION

Nancy Fraser's now classic 1996 essay brought the debate on justice and recognition to new heights. She reconceptualized much of the above discussion by arguing that claims for justice can be organized on two major structural axes—distribution and recognition—that constantly interact, but are not reducible to one another. Within each axis, she added, approaches to justice range between "affirmative" and "transformative" measures. Affirmative measures denote relatively cosmetic steps with a temporary effect on injustices, which tend to reproduce in the long-run the unequal capitalist/nationalist and male-dominated settings. Transformative measures, on the other hand, have more profound effects by challenging the social systems that produce the hierarchical order of classes, genders, "races," and ethnic entities. Fraser's intervention and the debates that ensued (see Young 1997; Fraser 2003) further entrenched recognition as a major category in the pursuit of social and urban justice (see Sandercock 2003).

Fraser's work included a profound critique of mainstream liberalism and of the increasingly popular procedural approaches to social justice (Fraser 1996). Her work had another effect—a welcome return to (refined) structuralism following a period in which Western theoretical debates were dominated by postmodernism and post-structuralism (see Soja 1995; Dear and Flusty 1998, 2002; Huxley and Yiftachel 2000). Postmodernism has also profoundly influenced planning, primarily through an apparent "communicative turn" that steered leading scholars (e.g., Forester 1993; Hillier 1993; Innes 1995; Healey 1997) towards micro-level investigations of the communicative (and chiefly procedural) interaction of planners and their working environments. It focused on the pursuit of Habermasian-inspired "communicative action" as the key to just and effective "deliberative" planning at the expense of more structural, material, or critical approaches (for a critique, see Huxley and Yiftachel 2000).

As noted by Fainstein (2005), at that stage the various camps in urban and planning theory, apart from radical Marxian and libertarian voices, agreed that recognition of diversity must be included in any consideration of a Just City. Questions remain on the right manner to approach difference and incorporate its various aspects into the planning process, but it was agreed by nearly all theorists that supporting diversity is "good," thereby providing a "new orthodoxy" for planning theory (Fainstein 2005: 1).

Yet, and this is our main theoretical point, it appears as if recognition was adopted somewhat uncritically. For most Western scholars, especially those advocating communicative or liberal/procedural approaches to justice, recognition became a catch-all phrase for the act of including minority or weakened groups, allowing them a "voice" in the policy process. Recognition was to be accepted as the liberal or civil "right" to be heard, to be counted and represented. Beyond a general support of inclusion and participation, we wish to advance three main lines of critique to this approach. First, recognition as a "right" presupposes a benign state and political setting and an operating constitutional democracy where rights can be secured through an independent judiciary. As observed by Fainstein (2005), who draws on Nussbaum (2002), rights alone are not enough and should be supplemented by "capabilities" in order to progress towards a Just City. Second, the emphasis and operationalization of liberal recognition is chiefly procedural; that is, focusing on participation and inclusion, but paying little attention to the material, economic, and concrete power aspects of planning recognition. There have been numerous accounts of this "thin" type of recognition, which often neglects—and is therefore blind to—material inequalities and oppressions (see McLoughlin 1992; Marcuse 2000).

Third and most importantly, liberal multicultural recognition tends to overlook the possibility that the marking of distinct groups may also harbor a range of negative consequences beyond the neglect implied by the previous point. As shown by various studies dealing with minorities, recognition may lead to a process of "othering" and may bear distinctively unjust material and political consequences. In other words, the institutional and legal "tagging" of a group as distinct, without strong civil constitutional foundations, may lead to outcomes very different from the inclusion and democratization sought by liberal scholars (see Samaddar 2005). This negative potential often surfaces in situations of ethnic, national, religious, and racial conflict, in which dominant groups are keen to reinforce the difference of weakened groups in order to perpetuate their disempowerment. The ethnic and racial élites may build on the existence of "deep difference" and "use" formal recognition to enable an ongoing process of exploitation and dispossession, now ostensibly achieved by an "inclusive process" (see Howitt 1998; Watson 2006).

URBAN NEOCOLONIALISM

The main point behind the need to reconceptualize recognition is the growing evidence of emerging urban neocolonial relations, which put in motion a pervasive process we define as "creeping apartheid" and the widespread emergence of "gray" (informal) space as part of today's urbanity. Urban colonialism sees dominant élites, whose privilege draws upon their identity, class, and location, utilize the contemporary city to advance three main dimensions of colonial relations: (1) expansion (of material or power position); (2) exploitation (of labor and/or resources); and (3) segregation (construction of hierarchical and essentialized difference).

To be sure, these dimensions operate today in geopolitical conditions very different from classical European colonialism. Most strikingly, the global European conquest and settlement is now reversed, with a flow of disenfranchised, often statusless, immigrants and indigenous peoples into the world's major cities. The economic power of the urban élites and the weakness and deep difference of immigrants (whether from rural regions or overseas) create patterns of ethno-class segregation and economic disparities that often resemble the traditional colonial city (King 1990; see al-Sayyad 1996). This urban order is most prevalent in liberalizing, ethnocratic states, which structurally privilege particular identities, while marginalizing minorities through both identity and economic régimes (see Law-Yone and Kalus 2001; Roy 2007; Tzfadia 2008).

These colonial-type urban relations are linked to the condition we term "creeping apartheid," in which groups enjoy vastly differing packages of rights and capabilities under the same urban régime, drawing on their class, identity, and place of residence. The order is "creeping" because it is never declared and is only partially institutionalized. Profound discrimination and inequality are based on both *de jure* and *de facto* mechanisms, which are commonly identified as "temporary." One of the most conspicuous "temporary" phenomena is the emergence of "gray" spaces composed of informal, often illegal, development and populations (see Roy and al-Sayyad 2004). Most typically, indigenous and immigrant minorities, squeezed between the various state and identity regulatory mechanisms, occupy and develop these gray spaces into a major component of today's metropolis, thereby augmenting the entrenchment of "creeping apartheid" (Yiftachel and Yacobi 2004). Hence, despite its putative "temporariness," this exploitative and uneven urban order has been intensifying for decades, and the population of disenfranchised urban residents and workers has grown significantly, often into the millions (Roy 2005; Davis 2006). In some cases, such as Dubai, Lagos, and Lima, the informal population has even become an urban majority (Davis 2006).

A variety of urban colonial relations are recorded by recent works emerging from, or about, non-Western cities, in which the majority of the world's urban population now resides (see Perera 2002; Robinson 2006). But this urban order, needless to say, is not confined to the global South, and is increasingly traced in cities of the developed world, mainly in focal points of mass immigration and economic growth, such as Paris and Los Angeles (see Marcuse 2000; Sassen 2006). Most of these studies find that identity and class inequalities are frequently connected, and that consequently, recognition and distribution intertwine in claims of social and spatial justice. Yet, identity and class also present different bases for human organization, which may undermine one another in the process of political mobilization. Hence they are not reducible to one another, requiring a more sophisticated treatment of policies allocating spatial resources, as does planning. As shown by Watson (2006), Robinson (2006), Huxley (2007), and Roy (2007), the role of space is pivotal, as both identity and class are actively shaped and reshaped through the ongoing production of urban space (e.g., high-rise development, suburbia, ghettoization, public housing, and immigration restrictions).

Given the above, we claim that the rubric of "identity," "diversity," "difference," and the catch-all "multiculturalism" are often too vague and at times confused in the current urban literature. We offer a conceptual way forward by sketching a continuum of recognition types, with three main "ideal types:" affirmative, indifferent, and hostile. These can assist in a more systematic analysis of the interaction between policy and identity.

Affirmative recognition entails recognition of a group's identity with the associated cultural and material needs and aspirations; allocation of a fair share of power and resources. There are two main sub-types: proportional and privileged recognition, reflecting the group's power and importance in the policy arenas. Affirmative recognition often leads to the constitution of amicable multicultural relations and inter-group integration in the city, although it may cause some tension with marginalized minorities, who may object to the advantageous position of privileged groups.

Indifference means the passive existence of the distinct group in the policy process. It entails non-recognition of the group's specific identity and its associated needs and demands, with official acceptance of its members as formally equal members of the urban community. Indifference leads to implicit and covert types of group domination and discrimination, deriving from the inability of minorities to pinpoint their discrimination in the absence of clear categories about their existence as a group. This often prevents them from setting legitimate collective goals. Sub-types include benign and marginalizing indifference, the first being typical of liberal régimes, where the promise of individual mobility tempers group grievances, while the latter typifies illiberal conditions, where group assimilation is coerced without strong commitment to civil rights. The consequences depend on specific geopolitical and economic conditions, although in general conflict levels are relatively low. The main focus of urban politics revolves around class and place, while identity politics is nudged to the periphery of the policy process.

Hostile recognition means the acknowledgment of group identity in policy-making with a concurrent framing of its demands in a range of negative images to the dominant perception of a good city. Hostile recognition constructs the group in question as a nuisance or threat. Subtypes vary between implicit and explicit hostility, which in turn fluctuates according to the nature of the groups in question. The consequences of hostile recognition also vary according to the group type, size, and setting, but they commonly cause the emergence of "gray" spaces of informal development and generate a dynamic of antagonism and polarization. Levels of conflict are highest when national or religious minorities with strong historical claims to the city are subject to this type of policy.

Notably, the above categories, and those used later in the essay, provide an analytical grid which cannot capture the complexity of the policy-recognition nexus. We suggest here a conceptual map to help discern and organize the complex field, with full awareness that all categories are socially constructed and are never stable or complete. The application of each type of recognition depends on a range of historical and political factors negotiated and determined in a wide range of societal spheres and struggles. They also depend on the variegated nature of group identities, which vary in their depth and future goals, ranging between separation, autonomy, integration, and assimilation.

Within this context, it is vital to remember that spatial policy is not a mere reflector of political forces imported from the "outside," but an important actor itself, which determines much of the way groups are treated in the public arena. While clearly set within an active political sphere, urban policy can assist in changing group position from marginalization and hostility towards recognition and equality, and vice versa, as depicted in Figure 7.1.

With this conceptual framework in mind, let us proceed to the planning of the Beer Sheva region and examine the ability of this framework to shed light on the connection between planning, justice, and the city.

PLANNING AND RECOGNITION(S) IN BEER SHEVA

Beer Sheva (Be'er Sheva' and Bi'r Saba'a in Hebrew and Arabic, respectively) is the main urban center of the Negev/Naqab region and a city of significant mythological and religious importance. It appears early in the Bible as the first town settled in the "promised land" by Abraham—the mythical father of Jews and Muslims. Today Beer Sheva accommodates a population of 186,000 in the city and some 560,000 in the metropolitan area (BSCC 2007). The modern city was rebuilt by the Ottoman Empire as an urban service and control center for surrounding Bedouin tribes, and this function continued during the British Mandate period, remaining a small and predominantly Arab town (Luz 2008).

Like other parts of Israel/Palestine, Beer Sheva became embroiled in the Zionist–Palestinian conflict. The 1947 UN partition plan included it under

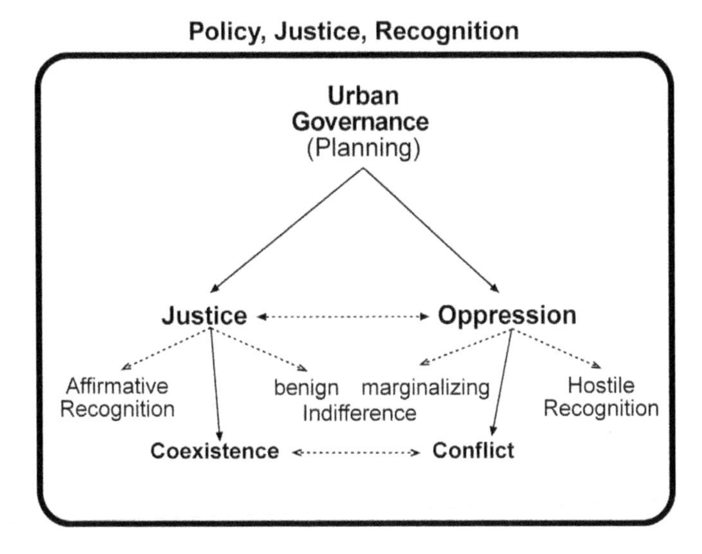

Figure 7.1 Policy, recognition, and justice

future Palestinian sovereignty, but the city was captured by Israel, which drove out its Arab population. During the 1948–49 war, some 70 to 80 percent of Arabs of the Naqab region were forced to leave, mainly going to Gaza, Egypt, the West Bank, and Jordan. Those 11,000 who remained were awarded Israeli citizenship, but were concentrated in a special military controlled zone known as "the siyag" ("the limit" or "fence," in Arabic "siyyaj") as depicted in Figure 7.2.

The ensuing decades saw the first wave of concerted Israeli effort to Judaize the previously Arab Naqab, using a combination of deeply ethnocratic land, development, housing, and planning policies. Israel nationalized nearly all Bedouin land (leaving about 15 percent of the region still under legal dispute), built eight new Jewish towns and some 105 rural Jewish settlements (see Falah 1983; Kedar 2003; Meir 2005). Masses of Jewish refugees and immigrants—mainly Mizrahim ("Eastern Jews") fleeing a hostile Arab

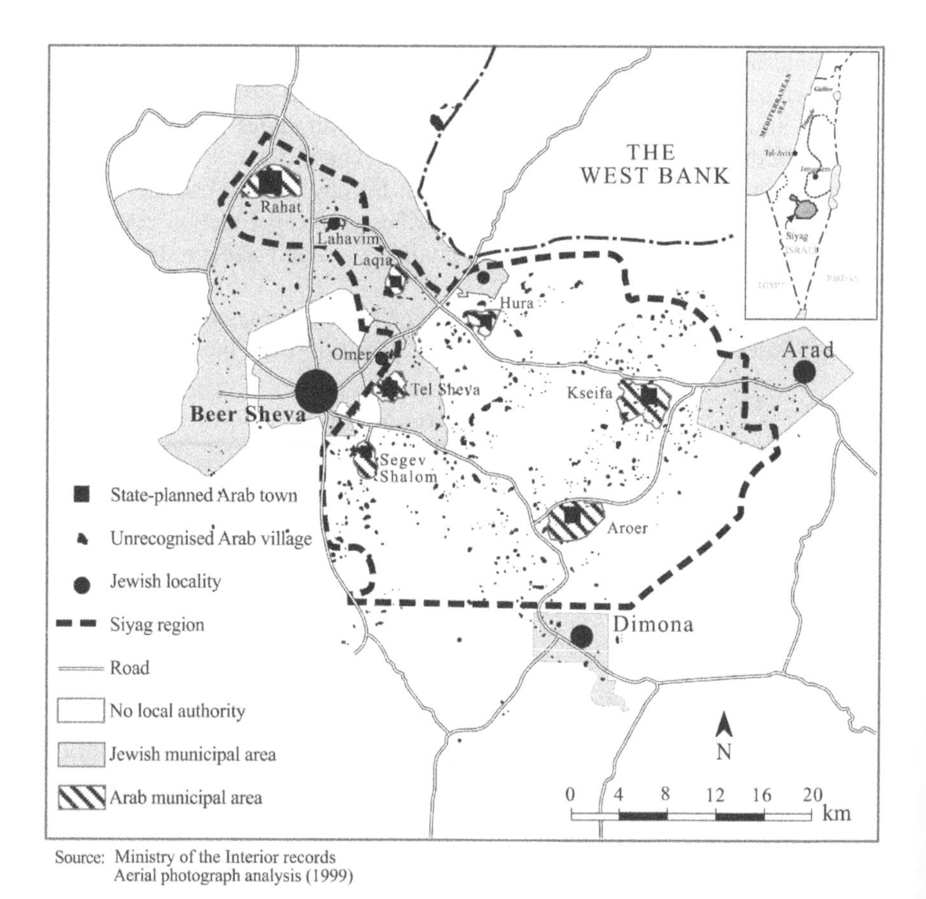

Figure 7.2 Jews and Arabs in the Beer Sheva metropolitan region

world—were housed in large public housing estates that were portrayed in the state planning discourses as the "national frontier" (Law-Yone and Kalus 2001).

In a few short years, however, the frontier, including Beer Sheva, turned into a marginalized periphery, in what was termed the "frontiphery" process (Yiftachel 2006). Subsequently, the Beer Sheva region became characterized by social and economic underdevelopment, mediocre levels of education and health, and a stigma deriving from its Mizrahi (Eastern) character (see Meir 2004; Cohen 2006). This was most conspicuous in the "development towns" —Israel's version of new town policy aimed at housing immigrants and creating new urban communities.

Eight such towns were built in the Beer Sheva region during the implementation of one of Israel's most ambitious planning projects. The towns housed large numbers of Mizrahim during the 1950s and 1960s, creating what Gradus and Stern (1980) called a southern "regiopolis." Small groups of immigrants continued to arrive during the 1970s and 1980s, mainly from the Soviet Union, France, and South and North America, although they did not significantly alter the region's Mizrahi character.

The next dramatic change occurred during the 1990s with a massive influx of mainly Russian-speaking immigrants from the former Soviet Union (hereafter "Russians") and some groups of Ethiopians. The city of Beer Sheva, like other localities in the region, welcomed the new influx, which facilitated large-scale development for accommodating the new housing demand, and adopted a new planning and public discourse of a "globalizing city" (see Gradus 2008; Markovich and Urieli 2008). This demand used the vast reserves of low-value state land, relaxed planning controls, and generous state incentives for large scale housing developments (see Alterman 1999; Tzfadia and Yacobi 2007; Shadar 2008).

The last wave of immigration resulted in a new ethnic composition. In 2007 the city population was composed of Mizrahim (41 percent), Russians (31 percent), Ashkenazim (8 percent), Ethiopians (4 percent), Arabs (3 percent), and six other small groups. In the wider metropolitan region, Mizrahim also constitute the largest group (29 percent), while Russians (24 percent) and Bedouin-Arabs (27 percent) also hold substantial proportions. The other groups are all smaller than 4 percent (NCRD 2007).

APPROACH

For this project we have attempted to analyze the overall impact of spatial policies on the main cultural groups in the Beer Sheva metropolitan region. To this end, we analyzed the plans affecting the city and region, which include: the 1952 national outline plan (later known as TAMA 1); the 1978 southern district plan (Plan 4/1); the 1991 national plan (TAMA 31); the 1996 development plan for Beer Sheva (non-statutory); the 1998 metropolitan plan for Beer Sheva region (Plan 4-14); the 2005 national plan (TAMA 35); the 2007 metropolitan plan (Amendment Plan 4-14-23); and supporting urban housing, land, and cultural policies of the Beer Sheva City Council.

These plans were developed by the ministries of Housing, Interior, and Infrastructure, and the Israel Land Authority, and they all attempt to create a major (Jewish) regional center in Beer Sheva. As noted by Gradus (1993), the above efforts have been only partially successful. Though the (Jewish) Beer Sheva region has become central to national and regional planning debates, which have led to significant new development, it has remained a peripheral urban region in terms of its economic, political, and cultural standing within Israeli/Palestinian space.

During this period of debate over the region, Israel implemented an urbanization planning strategy for the region's Bedouin Arabs. This has involved an attempt to concentrate the Bedouins into seven modern towns surrounding, but not part of, Jewish Beer Sheva (see Figure 7.2). This policy relocated about half the Arabs of the south (some 85,000 in 2007), mainly those with no land claims, through the lure of modern infrastructure and prospects of modernization. However, despite some development, the towns became known for their marginality, unemployment, deprivation, and crime (Yiftachel and Yacobi 2002; abu-Saad 2003). The remaining Bedouins, estimated at 80–90,000, have steadfastly stayed on their disputed land in some 45 unrecognized (shanty) towns and villages (Figure 7.2). A protracted land dispute over this "gray" space has persisted for decades.

The combination of these plans and policies as well as the accompanying discourses, regulations, and development initiatives are the subject of our analysis. We focus mainly on local and district plans and pay special attention to the implications of these plans for the region's main ethnic communities: Russian, Mizrahi, and Arab. We gain further insights by conducting a series of interviews with six key policy makers in the region, as well as 11 in-depth interviews with members of the communities in question.

PLANNING AND AFFIRMATIVE RECOGNITION: "RUSSIAN" IMMIGRANTS

Planning for immigrants from the former USSR in Beer Sheva has generally been marked by a benign attitude, premised on generous distribution and affirmative recognition, and couched within a long-term expectation of Russian integration into the Israeli-Jewish culture and society. The policy has been promoted jointly by an active state government and by urban authorities interested in accommodating the immigrants. The role of the Beer Sheva municipality has been central to this policy by actively luring these new immigrants, while some powerful urban centers in Israel, such as Tel-Aviv, Ra'anana, Ramat Gan, and Rishon Letzion, were indifferent and at times antagonistic to their entry (Tzfadia and Yacobi 2007).

The main thrust of urban policy towards the Russians, as reflected in National Plan No. 31 and the various Beer Sheva development plans, was the rapid provision of housing, first temporary and then permanent (see Alterman 2002; Gradus 2008). In parallel, the Israeli housing and planning systems thoroughly reorganized themselves and sped up the approval process, released previously protected agricultural land for urban development and provided generous subsidies and incentives for both immigrants and developers. A level of 65 percent home ownership was achieved in 2005, a mere 10–15 years after their mass arrival with meager financial or property resources (Tzfadia 2004).

The influx of over 40,000 Russian immigrants to Beer Sheva during the 1990s, and a corresponding period of rapid economic growth, spawned large-scale new housing and office construction. To illustrate, during 1989–2006 the city's population rose by 67 percent and the number of dwelling units rose by 86 percent, while office space increased by 51 percent (BSCC 2007). This caused large-scale residential relocations and significant vacancy chains, with many veteran residents upgrading their housing due to the availability of new government-assisted projects. Some housing vacancies were filled by the new immigrants, although most preferred to purchase new apartments due to the nature of government subsidies, which privileged new construction over existing housing stock.

Initially, the mass arrival caused economic and social concerns, because the population was relatively old and relied heavily on the city's welfare services. However, within a decade the economic benefits to the city outweighed the social costs, as the combination of social benefits and human skills propelled large sections within the Russian communities into the city's middle classes (Alias and Chaburstianov 2008; Gradus and Meir-Glitzenstein 2008).

City planning revisions created three large new neighborhoods on the outskirts of the city: Ramot, Nahal A'shan, and Neve Ze'ev/Nahal Beka (Figure 7.3). The latter two are characterized by their high percentage of Russians and their predominance in shaping the local landscape and institutions. The city council looked at the phenomenon favorably, as noted in an interview by Tal El-Al, a city councilor and member of the planning committee:

> The Russians are a blessing to our city; it's true that some needy population and social problems have arrived with them, but in the main they are a major asset to the city of Beer Sheva: educated, powerful, and urban . . . the city will continue to do all it can to absorb them in the best possible manner.

With regard to culture, state and city authorities, as well as market forces, combined to make a strong imprint on the urban environment in attempts to accommodate Russian immigrants. Large parts of Beer Sheva's urban landscape have been "Russified," with signs, institutions, and businesses catering to their growing demand for Russian products (especially food, drink, and sex). Alias and Chaburstianov (2008) surveyed the Russian cultural scene in Beer Sheva and discovered 11 bookstores, nine libraries, active theaters and community halls, and afternoon schooling in Russian. A Russian cultural enclave has been created, with Beer Sheva authorities providing support, finance, and planning.

This has also been reflected in Russian political organization, which formed several local parties—often associated with Russian state-parties. These created conspicuous levels of collective Russian political representation in City Hall, ranging between three to six seats in the city's 25-seat council. At the same time, Russians were increasingly appointed to professional positions with the City Council, culminating in the 2006 appointment of a Russian City Engineer and a Russian Chair of Planning services.

It is important to frame the Russian enclave—and the benign and affirmative recognition extended to that community—within the larger Israeli/Palestinian context. Russians are still expected by the majority of Israelis to integrate and eventually assimilate into the mainstream Jewish community. Israel has not adopted an open multicultural approach and denies the rights to separate Russian-language education and to separate legislation or institutions for autonomous governance. Partial Russian autonomy is created "from below" by communities, markets, and local governments, and this cultural

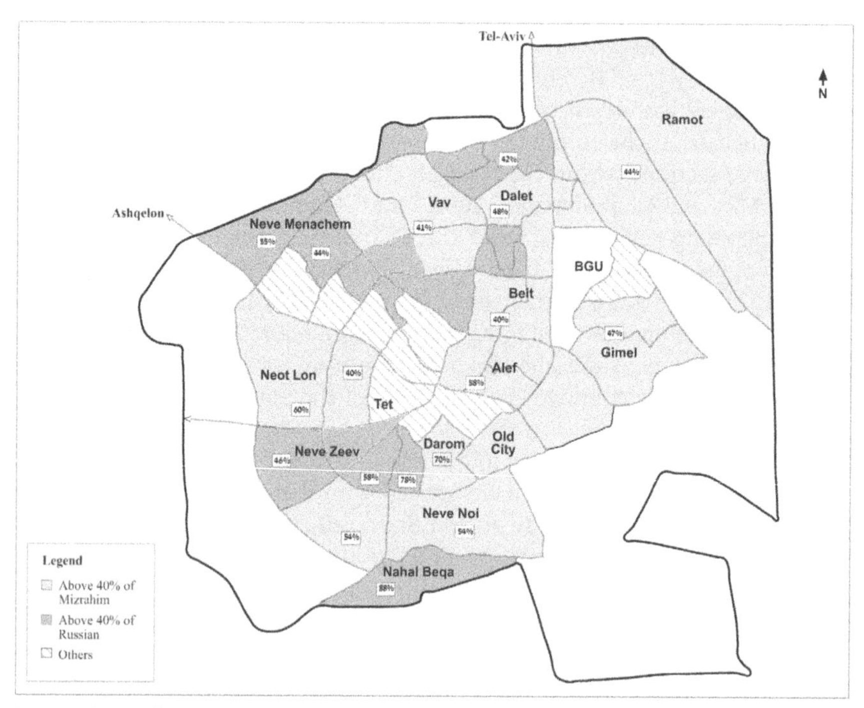

Source: Central Bureau of Statistics (Israel). Demographic characteristics of populations in cities and statistical areas: Population and households - select data. General Census 1999

Figure 7.3 Main ethnic areas in the city of Beer Sheva

autonomy is thriving due to the population's overall development in accordance with the Zionist state and its Judaization project. Even the large segments of this community who are not religiously Jewish (an issue of some concern in Israeli-Jewish circles) are also expected to integrate into the Israeli-Hebrew culture in a process termed elsewhere "the ethnicization of Zionism" (see Lustick 1999; Yiftachel 2006: chapter 5).

Yet, in planning terms, one can conceptualize the treatment of Russian immigrants as "a light side of planning," showing the ability of urban policies to effectively combine material distribution and benign cultural recognition and to extend justice-oriented policies towards an incoming low-income population. This is colorfully articulated by local Russian-Israeli poet, Victoria Orti:

> We came with fear and hope. The new country had a mystique, but also mountains and rivers of difficulties we had to cross. It appears like, after 15 years, that the mountains have lowered and the rivers have become shallower, and we are here—a struggling, but also thriving Russian community, with our identity, places and neighborhoods. We are well entrenched in our own state. We are here to stay.

PLANNING AND (MARGINALIZING) INDIFFERENCE: THE MIZRAHIM

The backbone of Beer Sheva's population is made up of the Mizrahim (Eastern Jews) who arrived *en masse* to the region during the 1950s and 1960s. The treatment of these migrants by urban authorities can be termed "marginalizing indifference."

From the outset, the Mizrahim were the "step-children" of Zionism (Shohat 2001), having been mobilized to join the Jewish national movement after the horrific consequences of the Nazi holocaust of the 1940s. As Zionist–Palestinian tensions rose, Arab régimes and Islamic societies became increasingly hostile to Middle Eastern Jewry, causing a mass exodus during the late 1940s and early 1950s (Behar 2007). Most of these Jews arrived in Israel and were housed by the state, first in temporary camps and later in mostly peripheral urban centers. Beer Sheva was one of the largest centers to accommodate Mizrahi immigration, with the city population rising six-fold between 1950 and 1970.

But the type of recognition extended to the Mizrahim was condescending and marginalizing. Their inclusion into the Zionist project was premised on their Judaism, but at the same time on a denial of their Eastern and Arab cultural affiliation. The state attempted to rebuild Jewish identity in the vision set by European secular élites. To that end, the masses of Mizrahim, who became a majority among Israel's Jews in the mid-1950s, had to be westernized, secularized, and de-Arabized (Swirski 1989; Shenhav 2006).

In Beer Sheva, as noted, Mizrahim quickly made a decisive majority, accounting for over 70 percent of the population. However, the city leadership remained predominantly Ashkenazi (western Jewish), headed by the "founding" and long-serving mayor of Israeli Beer Sheva, David Tuviyahu. The Ashkenazi–Mizrahi tension marked much of the local political scene during the first three decades of the state, but no genuine Mizrahi leadership could prevail at this time.

Over the years, Israel's national leadership has been keen to avoid severe ethnic riots in Beer Sheva, like those that earlier shook Haifa and Jerusalem. Therefore the Labor Party appointed a Mizrahi mayor, Eliyahu Nawy, who headed the city administration for nearly a decade during the 1970s. However, as noted by Meir (2004) and Cohen (2006), Nawy was selected precisely because he was a "soft" Mizrahi who could appease the masses in the predominantly Mizrahi city without threatening the state's Ashkenazi dominance in its peripheries. Following Nawy, and with the influx of Russian immigrants, the Mizrahi "threat" was blunted. The two long-serving mayors who followed, Yisrael Ragger and Yaakov Turner (the incumbent), came from the traditional Ashkenazi élites, preventing city Mizrahi communities from receiving open, public recognition.

Urban planning initiatives for the Mizrahi immigrants involved dense, modernist public housing developments located in a dozen centrally planned "garden city" type neighborhoods across the city (Figure 7.3). During the last two decades, several new, low-density neighborhoods and three suburban "satellite" towns have attracted most of Beer Sheva's (small) Ashkenazi population and those Mizrahim who moved into the middle classes. A degree of benign ethnic mixing began to occur in these localities, as it did in middle-class neighborhoods within the city limits. Large groups of Mizrahim still remain in the stigmatized inner city neighborhoods. Their employment was predominantly in labor-intensive industries and the low- to medium-level public sector, as well as in small trade and local business. This created a conspicuous overlap between their Mizrahi ethnicity and working and lower middle-class position (see Yonah and Saporta 2003).

Mizrahi local politics, which often reflected national trends, shifted over time from supporting the Labor Party, associated with early Zionism and state building, to the Rightist and nationalist Likud, and most recently to the Orthodox-Mizrahi Shas movement (Meir 2004; Tzfadia 2004). Critically for the issue of recognition, the organization of Mizrahi parties was constantly undermined by the state and city leadership and portrayed as "divisive" and harmful to the Israeli state project (see Peled 2001; Grinberg *et al.* 2005). This was reflected in the identity and activity of the city mayors, as noted above (see Cohen 2006; Meir-Glitzenstein 2008). The lack of Mizrahi political organization stands in contrast to other ethnic parties, such as those representing Russian and Orthodox Jews, and later Arabs, who were accepted as legitimate by the Israeli élites and public. The local Mizrahi majority did attempt to form a political block on several occasions during the 1960s, 1970s, and 1980s, but was continuously thwarted by the concerted campaign to delegitimize Mizrahi "divisive" mobilization (Meir 2004; Cohen 2006).

The lack of political recognition was mirrored in the cultural sphere. During Israel's early decades, Mizrahi culture was stigmatized in Israel and, by implication, in Beer Sheva. Most aspects of Mizrahi identity—family, dress, language, music, dwelling, and even religion—were silenced or ridiculed in the public discourse, education system, cinema, and popular culture (Shohat 2001). "Levantine" became synonymous with "primitive," leaving strong Mizrahi localities like Beer Sheva in a deep identity crisis. In Beer Sheva, due to the sheer size of the Mizrahi population, the public culture inevitably bore many Mizrahi features, but as Cohen (2006: 5) summarizes:

> The depth of Beer Sheva's stigma . . . derives directly from its open Levantine character. [T]he denial of the city's Levantine culture is central to Israel's Eurocentric discourse, which places European Jews at the center of the Zionist project . . . The self-portrayal of most Israelis as "Westerners" requires stigmatizing Beer Sheva by stressing negatively both its Arab past and a Mizrahi present.

But this critical self-observation remains on the periphery of the public debate, and the marginalizing indifference among the élites by and large continued until the late 1980s. During the last two decades, some change can be traced, termed earlier "multiculturalism from below." The more liberal attitudes of recent years has yielded a measure of cultural recognition, revolving around Mizrahi holidays, music, food, and cultural events, although these are more typically assigned to sub-groups (e.g., "Moroccan" or "Yemenite") than to a general Mizrahi identity. This liberalization serves to further highlight the lack of political organization and Mizrahi narrative in the urban public sphere. The city leadership continues to gloss over differences between Jewish ethnicities, as noted by city planner Offer Ilan:

> . . . the question of the city's Mizrahim has never fully arisen in planning and policy circles. We have other categories, such as "new immigrants," "low-income," "religious," and "ultra-orthodox." All of these are addressed by our planning and development strategies. But Mizrahim? I cannot see the relevance. I am a Mizrahi, too, and it has little to do with urban planning; it's time to understand that the Mizrahim have integrated into Israeli society, including Beer Sheva. Sure, they have problems, but not as Mizrahim.

PLANNING AND HOSTILE RECOGNITION: BEDOUIN ARABS

One central aspect of spatial policy in the Beer Sheva metropolitan region has been the hostile recognition extended to the region's Bedouin Arab community. A bitter land conflict has developed with the state, which has continuously denied the Bedouins indigenous land rights, and as a result declared them "invaders" to their own historic localities. In an effort to force them to relocate, the state has refused to recognize their land claims and has prevented the supply of most services, including roads, electricity, clinics, and planning. House demolition campaigns are launched on a regular basis (see Meir 2005; Yiftachel 2006).

Levels of poverty, child mortality, and crime are the worst in Israel/Palestine and create a metropolitan geography of stark ethno-class contrast to the well-serviced adjacent Jewish localities. The Beer Sheva metropolis has come to resemble many Third World cities that comprise a well-developed, modern urban core and a range of peripheral, informal localities suffering severe poverty and deprivation. It is here that the process of urban colonialism and "creeping apartheid" noted above are most evident.

Expressions of urban colonialism have therefore been pervasive in the Beer Sheva region, but less confrontational than in internationally known cases like Hebron and Jerusalem. Arab campaigns against deprivation have highlighted both equality and identity, focusing on the right to reasonable material conditions as well as cultural preservation (see Meir 2005). In recent years, religion has played an increasing part in Arab urban campaigns, especially around education and places of worship.

Bedouin Arab representation in urban and regional planning affairs has ranged between non-existent and negligible. Despite being the indigenous inhabitants of the region and constituting nearly a third of its current population, Bedouin presence in planning bodies has been meager and random. During the last decade, for example, only two Bedouins have sat on the district planning council (each in turn being one amongst 13 Jews on the council), and not even one Bedouin is represented on the Beer Sheva city council. Other planning bodies, such as the Israel Land Authority and the Ministry of Housing, Welfare, and Education, have occasionally included a single Arab member, but always in a position of distinct minority.

The combination of land, cultural, and material deprivations and a lack of representation, has bolstered antagonism towards the state and spurred the Bedouin Arabs to form their own institutions. The Regional Council for the Unrecognized Villages (RCUV) was formed in 1997 to combine the various localities surrounding Beer Sheva and to present an alternative planning approach based on full recognition of indigenous rights and equality. This form of "insurgent planning" (see Meir 2005) rallied a group of notable NGOs to support the new (unrecognized) council and caused some change in the public discourse. It is no longer possible to ignore the Bedouins as mere "invaders" or "outsiders" to the metropolitan region, and their demands are heard continuously in the media and in administrative and professional circles.

The authorities have also been forced to recognize nine of the 45 unrecognized villages, although no infrastructure, such as running water, roads, and permanent schools, has been allocated to these localities yet. Insurgent indigenous planning practices and the prevailing attitude of "hostile recognition" has clashed in recent years to cause spiraling polarization between Bedouins and authorities, with little progress towards resolving the conflict seen (see Yiftachel 2006). One such issue revolves around the renowned and architecturally significant Beer Sheva mosque, which was built by the Ottomans to serve the region's population. Despite constant Arab demands, the city refuses to open it for Muslim worship, with one powerful councilor of the ruling coalition, Eli Bokker, claiming that "the region has dozens of mosques in Bedouin localities and towns, and Beer Sheva is now a Jewish city with the right to protect this urban character."[1]

As a result, the mosque has been lying idle for decades and is now in an advanced state of architectural deterioration. Following a recent appeal by several NGOs, the Israeli high court ruled in favor of opening the mosque for "Arab cultural uses." Despite the latest ruling, the city is steadfast in its refusal and has now condemned the building as too dangerous for human use. Those against opening the mosque were members of the nationalist Likud Party and the (mainly Russian) Yisrael Beitenu and Mizrahi religious parties. Yaacov Margi, a Beer Sheva Shas leader claimed:

> This high court decision could be the last nail in the Beer Sheva coffin . . . [W]e have been increasingly surrounded by Bedouins from all sides, and they now attempt to penetrate the heart of our city by opening their mosque . . . Let us never forget: Beer Sheva is where Abraham's wells are still in existence after 4,000 years. We should continue to drink the wisdom of our Tora like the water from these wells and remember that one of these wisdoms is to never, but never, let the Amalek [hostile nations] raise their heads!

Margi's statement is a reminder of the powerful narratives framing urban colonialism and the resultant politics of denial, fear, and hostile recognition. In this process, termed "creeping apartheid," a distant mythical Jewish past and vague future (of possible immigration) are fully recognized, while the current needs of Moslem residents are actively denied.

IMPACT AND REFLECTION

The foregoing shows that, indeed, groups are recognized in very different ways by the urban policy process. One clear question that arises from this is what the long-term impact of such uneven recognition is, although its systematic examination must await a different context.

Still, it is not difficult to intuitively associate negative types of recognition with socioeconomic marginalization and political weakness. This is supported by a cursory look at the socioeconomic standing of urban communities in the Beer Sheva region. We can take, for example, the "quality of life" index of localities prepared by the Israel Bureau of Statistics, which is based on a combination of socioeconomic characteristics (CBS 2006). In the 2005 survey, the typical Russian neighborhood of Neve Ze'ev received a score of 12 (in a 1–20 range), while a decade earlier it received only a score of eight.

Another concentration of Russians, Nahal A'shan, received the score of nine against six a decade earlier.

Such change was not the case in typical Mizrahi neighborhoods, such as Schuna Gimmel, which received a score of eight in 2006 and nine in 1995, and Schuna Tet, a more middle-class neighborhood with large Mizrahi concentration, which had the same score of 13 in both years. The Bedouin Arab localities surrounding Beer Sheva, Tel Sheva, and Laqiyya scored three and four, respectively in 2005, and two and three a decade earlier. These scores indicate the significant improvement of localities identified with Russians, as opposed to the stagnation characterizing localities with Mizrahi and Arab majorities. They also highlight notable differences within each cultural group, indicating that other forces are at work in the stratification process. Needless to say, the link between recognition and development requires a more in-depth investigation. This cursory look confirms, however, the importance of considering the specific type of recognition as a key element in theorizing justice and oppression in the city.

Our understanding of spatial justice has indeed been complicated in recent years by the introduction of "recognition" as a major philosophical axis for justice claims and by the mobilization of politics of identity. Recognition claims interact in complex ways with the well-established call for fair distribution of material and political resources and fairness in decision-making processes. The nature of this interaction is further complicated by our main argument in this essay, namely, that recognition has to be studied critically and that it may work for or against the group in question.

Clearly, the questions raised in this essay present a major challenge to the justice literature and need to be explored further, both theoretically and empirically. The need for this investigation is reinforced in the rapidly changing urban world, where diversity, hierarchy, and identity politics are rewritten within a globalizing economy, and within new régimes of uneven citizenship. We plan to continue the current exploration both comparatively, examining various types of ethnically divided cities, and theoretically, engaging new debates over spatial justice which emerge from changing urban and political environments.

A promising, if understudied, way forward may be found in the further development of the Lefebvrian notion of "the right to the city." As Fainstein (2005) rightly notes, Lefebvre's work is highly abstract, lacking in specific details on the precise nature and applicability of this right. But perhaps precisely because of its abstraction, the main plank in Lefebvre's framework, the right to use and appropriate the city's main features of centrality and difference, can now be injected with new meanings, reflecting a need to extend benign forms of recognition to all groups residing in the city. In such settings, urban colonialism and "creeping apartheid" may be transformed into new forms of urban federalism based on equality, autonomy, and redistribution. The translation of this idealistic vision into urban policies appears to be an appropriate challenge for students of the future Just City.

NOTE

1 *Sheva* (local newspaper), May 9, 2005; a similar statement was made by Bokker in *Kolbi* (another local paper) on May 8, 1998.

REFERENCES

abu-Saad, I. (2003) Bedouin Towns in the Beginning of the 21 Century: Negev Bedouin Following the Failure of Urbanization Policy. *Sikkuy Report, 2002–2003*. Jerusalem and Tamra: Sikkuy; 49–59 (Hebrew and Arabic).

Alias, N. and N. Chaburstianov (2008) "Not on Bread Alone: the Cultural Life of the Beer Sheva Russian Street," in Y. Gradus and E. Meir-Glitzenstein (eds) *Beer Sheva: Metropolis in the Making*, Beer Sheva: Negev Center for Regional Development, Ben-Gurion University Press: 45–68 (Hebrew).

al-Sayyad, N. (1996) "Culture, Identity and Urbanism in a Changing World: a Historical Perspective on Colonialism, Nationalism and Globalization," in M. Cohen, B. Ruble, J. Tulchin and A. Garland (eds.) *Preparing for the Urban Future: Global Pressures and Local Forces*, Baltimore: Woodrow Wilson Center Press; 106–133.

Alterman, R. (1999) *Farm Land Between Privatisation and Continued National Ownership*, Jerusalem: Floresheimer Institute (Hebrew).

—— (2002) Planning in the Face of Crisis: Land, Housing and Immigration in Israel. London, Routledge.

Badcock, B. (1984) *Unfairly Structured Cities*, Oxford: Blackwell.

Behar, M. (2007) "Palestine, Arabized Jews and the Elusive Consequences of Jewish and Arab National Formations," *Nationalism and Ethnic Politics*, 13(4): 1353–7113.

Bollen, S. (1999) *Urban Peace-Building in Divided Societies*, Boulder: Westview Press.

—— (2007) *Cities, Nationalism, and Democratization*, Oxford and New York: Routledge.

BSCC (Beer Sheva City Council) (2007) *Annual Report*, Beer Sheva: City Council.

CBS (Central Bureau of Statistics, Israel) (2006) *Quality of Life Index 2005*, Jerusalem: CBS.

Castells, M. (1978) *The Urban Question*, London: Edward Arnold.

Cherry, G. (1988) *Cities and Plans*, London: Edward Arnold.

Cohen, E. (2006) *Beer Sheva—the Fourth City*, Jerusalem: Carmel.

Davis, M. (2006) *Planet of Slums*, London, Verso.

Dear, M. (1981) "Towards a Theory of the Local State," *Political Studies from a Spatial Perspective*, A. Burnett and P. Taylor. New York: Wiley and Sons.

Dear, M. and S. Flusty (1998) "Postmodern Urbanism," *Annals of the Association of American Geographers*, 88(1): 50–72.

—— and —— (eds) (2002) *The Spaces of Postmodernity: Readings in Human Geography*, London: Blackwell.

Fainstein, S. (2005) "Cities and Diversity: Should We Want It? Can We Plan for It?," *Urban Affairs Review*, 41; 1: 3–19.

Falah, G. (1983) "The Development of Planned Bedouin Resettlement in Israel, 1964–82: Evaluation and Characteristics," *Geoforum*, 14: 311–323.

—— (1989) "Israelisation of Palestine Human Geography," *Progress in Human Geography*, 13: 535–550.

Forester, J. (1993) *Critical Theory, Public Policy and Planning Practice: Toward a Critical Pragmatism*, Albany: State University of New York Press.

Fraser, N. (1996) "Recognition or Redistribution? A Critical Reading of Iris Young's Justice and the Politics of Difference." *Journal of Political Philosophy*, 3 June: 166–80.

—— (2003) *Redistribution or Recognition: A political-philosophical exchange*, New York: Verso.

Friedmann, J. (2002) *The Prospects for Cities*, New York: University of Minnesota Press.

Gradus, Y. (1993) "Beer-Sheva—Capital of the Negev Desert," in Y. Golani, S. Eldor and M. Garon (eds.) *Planning and Housing in Israel in the Wake of Rapid Changes*. Jerusalem, The Ministry of the Interior: 251–65.

—— (2008) "The Beer Sheva Metropolis: Polarized Multicultural Urban Space in the Era of Globalization," in Gradus, Y. and E. Meir-Glitzenstein (eds) *Beer Sheva: Metropolis in the Making*. Beer Sheva, Negev Center for Regional Development, Ben-Gurion University Press (Hebrew).

Gradus, Y. and E. Meir-Glitzenstein (eds) (2008) *Beer Sheva: Metropolis in the Making*. Beer Sheva: Negev Center for Regional Development, Ben-Gurion University Press (Hebrew).

Gradus, Y. and Stern, E. (1980) "Changing Strategies of Development: Toward a Regiopolis in the Negev Desert," *Journal of the American Planning Association*, 46: 410–423.

Grinberg, L., Abutbul, G. and Mutzafi-Hellar, P. (2005) *Mizrahi Voices: Towards a New Mizrahi Discourse on Israeli Society*, Tel-Aviv: Massad (Hebrew).

Hague, C. (1984) *The Development of Planning Thought*, London: Hutchinson.

Hall, P. (1988) *Cities of Tomorrow*, Berkeley: Basil Blackwell.

Hall, S. (1991) "Old and New Identities, Old and New Ethnicities," in King, A. (ed.) *Culture, Globalization and the World System*, London: MacMillan.

Harvey, D. (1973) *Social Justice and the City*, London: Edward Arnold.

Healey, P. (1997) *Collaborative Planning: Shaping Places in Fragmented Societies*, London, Macmillan.

Hillier, J. (1993) "To boldly go where no planners have ever . . ." *Environment and Planning D*, 11(1): 89–113.

hooks, b. (1995) *Killing Rage: ending racism*, New York: Macmillan.

Howitt, R. (1998) "Recognition, Respect and Reconciliation: Steps towards Decolonisation?" *Australian Aboriginal Studies*, 16(2): 3–16.

Huxley, M. (2007) "Geographies of governmentality," in J. Crampton and S. Elden (eds) *Space, Knowledge, Power: Foucault and Geography*, London: Ashgate; 87–109.

Huxley, M. and O. Yiftachel (2000) "New Paradigm of Old Myopia? Unsettling the Communicative Turn in Planning Theory," *Journal of Planning Education and Research*, 19(4): 333–342.

Innes, J. (1995) "Planning Theory's Emerging Paradigm: Communicative Action and Interactive Practice," *Journal of Planning Education and Research*, 14(3): 183–191.

—— (1995) "Planning Theory's Emerging Paradigm: Communicative Action and Interactive Practice," *Journal of Planning Education and Research*, 14(3): 183–191.

Jacobs, J. (1996) *Edge of Empire*, London: Routledge.

Kedar, S. (2003) "On the Legal Geography of Ethnocratic Settler States: Notes Towards a Research Agenda," in J. Holder and C. Harrison (eds.) *Law and Geography Current Legal Issues*. Oxford: Oxford University Press; 401–442.

King, A. (1990) *Urbanism, Colonialism and the World Economy*, London: Routledge.
Kymlicka, W. (1995) *Multicultural Citizenship: a Liberal Theory of Minority Rights*, Oxford: Clarendon Press.
Law-Yone, H. and Kalus, R. (2001) "The Dynamics of Ethnic Spatial Segregation in Israel," in O. Yiftachel (ed.) *The Power of Planning: Spaces of Control and Transformations*, The Hague: Kluwer Academic.
Lefebvre, H. (1996) "Philosophy of the City and Planning Ideology," *Writings on Cities*, London: Blackwell; 97–101.
—— (1991) *The Production of Space*, Oxford: Blackwell.
Lustick, I. (1999) "Israel as a Non-Arab State: the Political Implications of Mass Immigration of Non-Jews," *Middle East Journal*, 53(3): 417–433.
Luz, N. (2008) "The Making of Modern Beer Sheva—an Imperial Othoman Project," in Y. Gradus and E. Meir-Glitzenstein (eds) *Beer Sheva: Metropolis in the Making*, Beer Sheva: Negev Center for Regional Development, Ben-Gurion University Press: 161–174 (Hebrew).
Marcuse, P. (1978) "Housing Policy and the Myth of the Benevolent State," *Social Policy*, (January/February): 21–26.
—— (2000) "Identity, Territoriality and Power," *Hagar: International Social Science Review* 1(1): 128–143.
Markovitz, F. and N. Urieli (2008) "Consumerism and Global/Local Identity in the Negev: the 'BIG' Center and Beer Sheva's Old City," in Y. Gradus and E. Meir-Glitzenstein (eds) *Beer Sheva: Metropolis in the Making*, Beer Sheva: Negev Center for Regional Development; 212–228 (Hebrew).
Massey, D. (2007) *For Space*, London: Sage.
McLoughlin, J.B. (1992) *Shaping Melbourne's Future?: Town Planning, the State and Civil Society*. Cambridge: Cambridge University Press.
Meir, A. (2005) "Bedouins, the Israeli state and insurgent planning: Globalization, localization or glocalization?" *Cities*, 22(3): 201–235.
—— (2008) "Negev Bedouins, Globalization and Planning and Metropolitan Beer Sheva," in Y. Gradus and E. Meir-Glitzenstein (eds.) *Beer Sheva: Metropolis in the Making*, Beer Sheva: Negev Center for Regional Development, Ben-Gurion University Press: 81–106 (Hebrew).
Meir, E. (2004) "Zionist and Arab-Jewish Identity in the Collective Memory of Iraqi Jews in Israel." *Alpayim*, 27: 44–70 (Hebrew).
Meir-Glitzenstein, E. (2008) "The Ethnic Struggle in Beer Sheva During the 1950s and 1960s," in Y. Gradus and E. Meir-Glitzenstein (eds) *Beer Sheva: Metropolis in the Making*, Beer Sheva: Negev Center for Regional Development; 23–44 (Hebrew).
Mitchell, D. (2003) *The Right for the City: Social Justice and the Fight for Public Space*, New York: Guilford.
NCRD (Negev Center for Regional Development) (2007) *Negev Statistical Year Book*, Beer Sheva, Negev Center (Hebrew).
Nussbaum, M.C. (2002) *Beyond the Social Contract: Capabilities and Global Justice*, New Delhi: Oxford University Press.
Peled, Y. (ed.) (2002) *Shas: the Challenge of Israeliness*, Tel-Aviv: Yediot Ahronot (Hebrew).
Perera, N. (2002) "Indiginising the Colonial City: Late 19th-century Colombo and its Landscape," *Urban Studies*, 39(9): 1703–1721.
Robinson, J. (2006) *Ordinary Cities: between Globalization and Modernity*, London: Routledge.

Roy, A. (2005) "Urban Informality: Toward an Epistemology of Planning," *Journal of the American Planning Association*, 71(2): 147–158.
—— (2007) "The 21st Century Metropolis: New Geographies of Theory," *Regional Studies* 41.
Roy, A., al Sayyad, N. (eds) (2004) *Urban Informality in the Era of Globalization: A Transnational Perspective*, Boulder: Lexington.
Samaddar, R. (2005) *The Politics of Autonomy: Indian Experiences*, New Delhi: Sage.
Sandercock, L. (1995) "Voices from the Borderlands: a Mediation of a Metaphor," *Journal of Planning Education and Research* 14: 77–88.
—— (1998) *Toward Cosmopolis: Planning for Multicultural Cities and Regions*, London: Wiley and Sons.
—— (2003) *Mongrel Cities of the 21st Century*, New York: Continuum Press.
Sassen, S. (2006) *Territory, Authority, Rights: From Medieval to Global Assemblages*, Princeton: Princeton University Press.
Shadar, H. (2008) "Ideologies in the Planning of Beer Sheva," in Y. Gradus and E. Meir-Glitzenstein (eds) *Beer Sheva: Metropolis in the Making*, Beer Sheva: Negev Center for Regional Development, Ben-Gurion University Press: 175–199 (Hebrew).
Shenhav, Y. (2006) *The Arab Jew: a Postcolonial Reading of Nationalism, Religion, and Ethnicity*, Stanford: Stanford University Press.
Sohat, E. (2001) *Forbidden Reminiscences*, Tel-Aviv: Kedem Publishing.
Soja, E. (1995) "Heterotopologies: A Rememberance of Other Spaces in the Citadel-LA," S. Watson and K. Gibson (eds) *Postmodern Cities and Spaces*. Oxford: Basil Blackwell; 13–34.
Swirski, S. (1989) *Israel: the Oriental Majority*, London: Zed.
Taylor, C. (1992) "The Politics of Recognition," in A. Gutman (ed.) *Multiculturalism: Examining the Politics of Recognition*, Princeton, New Jersey: Princeton University Press; 25–73.
Thomas, H. (1995) "Race, Public Policy and Planning in Britain," *Planning Perspectives*, 10: 125–148.
Thomas, H. and Krishnarayan, V. (1993) "Race Equality and Planning," *The Planner*, 79(3): 17–21.
Troy, P. (ed.) (1981) *Equity in the City*, Sydney: George Allen and Unwin.
Tzfadia, E. (2004) " 'Trapped' Sense of Peripheral Place in Frontier Space," in H. Yacobi (ed.) *Constructing a Sense of Place—Architecture and the Zionist Discourse*, Burlington: Ashgate: 119–135.
—— (2008) "New Settlements in Metropolitan Beer Sheva: the Involvement of Settlement NGO's," in Y. Gradus and E. Meir-Glitzenstein (eds) *Beer Sheva: Metropolis in the Making*, Beer Sheva: Negev Center for Regional Development, Ben-Gurion University Press: 107–119.
Tzfadia, E. and Yacobi, H. (2007) "Identity, Migration, and the City: Russian Immigrants in Contested Urban Space in Israel," *Urban Geography*, 28(5): 436–455.
Watson, V. (2006) "Deep Difference: Diversity, Planning and Ethnics," *Planning Theory*, 5(1): 31–50.
Yiftachel, O. (1991) "State Policies, Land Control and an Ethnic Minority: the Arabs in the Galilee, Israel," *Society and Space*, 9: 329–362.
—— (2002) "Territory as the Kernel of the Nation: Space, Time and Nationalism in Israel/Palestine," *Geopolitics*, 7(3): 215–248.

—— (2006) *Ethnocracy: Land and Identity Politics in Israel/Palestine*, Philadelphia: Pennpress, University of Pennsylvania.

—— (2007) "Re-Engaging Planning Theory," *Planning Theory*, 5(3): 211–222.

Yiftachel, O. and H. Yacobi (2002) "Planning a Bi-National Capital: Should Jerusalem Remain United?," *Geoforum*, 33: 137–145.

—— and —— (2004) "Control, Resistance and Informality: Jews and Bedouin-Arabs in the Beer-Sheva Region," in N. Al-Sayyad and A. Roy (eds) *Urban Informality in the Era of Globalization: A Transnational Perspective*, Boulder: Lexington Books; 118–136.

Yonah, Y. and Y. Saporta (2003) "Land and Housing Policy in Israel: the Discourse of Citizenship and Its Limits," in Shenhav, Y. (ed.) *Space, Land, Home*, Jerusalem: Van Leer; 129–152.

Young, I.M. (1990) *Justice and the Politics of Difference*, Princeton, New Jersey: Princeton University Press.

—— (1997) *Intersecting Voices: Dilemmas of Gender, Political Philosophy and Policy*, Princeton, New Jersey: Princeton University Press.

8 On globalization, competition, and economic justice in cities

James DeFilippis

INTRODUCTION

In this chapter I discuss the issue of economic justice in cities, and do so in ways that attempt to refocus our attention away from economic globalization and towards forms of employment that are rooted in place. Focusing on the most exploitative and unjust jobs that currently exist in American cities, the chapter documents the ways in which these hyper-exploitative jobs are organized, and who the workers are that are employed in them. It emphasizes that the vast majority of these jobs are in sectors that are localized in their products, markets, and competition. Thus the economic injustices in cities are more often the result of class-based competition within cities than they are functions of any extra-local competition. Unfortunately, by emphasizing the wrong forms of competition some urban scholarship has enabled the state and capital to undermine not just efforts for social redistribution, but also the regulation of the social relations in work. It has therefore facilitated the growth of economic injustice in American cities. Fortunately, efforts for economic justice that take employment relations within cities as focal points of their work have been and are being organized as part of the struggle to realize more just cities. I conclude this chapter by discussing some of these efforts.

THE GLOBAL TROPE

The relationships between cities, competition, the global economy, and social and economic justice have received a great deal of attention in the last 20 years. In the vast literature that examines these relationships, it is usually assumed that the issue is competition between people, usually élites, in localities[1] who are fighting for investment of capital and the economic development that it theoretically would bring. This understanding of cities existing in the midst of global competition, what Wilson (2007) calls "the global trope," is one of the dominant mantras of the contemporary political economy.

This trope of cities in competition is usually how the issues of economic development in cities are framed and understood by planners and other urbanists—often through a lens of the "trade-offs" between economic efficiency and economic equity or justice. In some highly influential cases, both academically and in the public at large, the framing of the issue has been explicit: places cannot engage in efforts that limit capital accumulation, because if they do, capital will "vote with its feet" (Tiebout 1956; Peterson 1981) and flee such places. Thus efforts for modest redistribution, themselves far cries from trying to realize social or economic justice, are understood to be doomed to fail, because the capital needed for them will simply locate elsewhere. It would be easy to pull out many quotes that exemplify this understanding of economics and place, but one from Thomas Friedman—perhaps the best known and most publicly influential of the "There Is No Alternative" school in the United States—should suffice, and Friedman bluntly states: "If you don't run with the Global Herd and live by its rules, accept the fact that you are going to have less access to capital, less access to technology, and ultimately a lower standard of living" (Friedman 1999: 168).

This understanding of the issues is limiting—not just philosophically, but empirically, pragmatically, and politically. Frustration with how myopic this understanding is has been expressed well by Francis Fox Piven who argued that:

> The key fact of our historical moment is said to be the globalization of national [or, I might add, local] economies . . . I don't think this explanation is entirely wrong but it is deployed so sweepingly as to be misleading. And right or wrong, the explanation itself has become a political force, helping to create the institutional realities it purportedly merely describes.
>
> (Piven 1995: 107)

David Harvey made the same point rather well when he said:

> There is a strong predilection these days to regard the future of urbanization as already determined by the powers of globalization and of market competition. Urban possibilities are limited to mere competitive jockeying of individual cities for position within the urbanization process from which to launch any kind of militant particularism capable of grounding the drive for systemic transformation . . . The ideological effect of this discursive shift has been extraordinarily disempowering with respect to all forms of local, urban, and even national political action.
>
> (Harvey 1996: 420)

As Fainstein rightly argues (see Chapter 1 in this volume), the act of naming has power. So we need to be clear about how we understand and name the processes by which economic *injustice* is perpetuated in urban space. And the framing of the processes of *injustice* as somehow connected to issues of economic competition between places has, as Harvey and Piven have argued, significantly limited our political imagination. It has also shifted the focus away from urban economic injustices that are rooted in the much longer-term processes of capitalist urbanization and, I would argue, misunderstood the nature of cities in the global economy. Rightly understood, urban economic injustices are often much more localized sets of relations and issues. In short, we not only let off the hook those that perpetuate injustice by allowing them to hide behind "global competition" but we make our quest for justice much harder by focusing our aim on the wrong targets and using the wrong explanatory framework.

GLOBALIZATION AND THE LOCALIZATION OF EMPLOYMENT

One of the remarkable things about cities in the global north, including—or perhaps especially—even the most heralded global cities like New York, is that globalization, rather than simply making urban economies more global, has in many ways also made them more local. This is because, while cities have become focal points for whole sets of relations that are transnational, they have simultaneously seen massive growth in the components of their economies that are localized in their products, inputs, and markets.

If you stay with the example of New York City, and consider the composition of the city's employment, the numbers are striking. The city has gone from employing one million people in manufacturing in 1950—most of it for export, and exposed to extra-local competition—to employing only 115,000 in 2005 (Bernhardt 2006). The related processes of capital mobility, economic restructuring, and globalization have already long-since driven most of the manufacturing jobs exposed to global competition out of the city (although it is certainly true that the decline in the city's manufacturing base has also been a product of re-zoning manufacturing land to other uses, and the blame for local deindustrialization cannot be placed solely at the feet of global competition; see Fitch 1993; Pratt Center for Community and Environmental Development 2001). While the finance, insurance, and real estate (FIRE) sector has usually been assumed to pick up for manufacturing, the total number of finance-sector jobs in 2005 was 450,000—which is the same as it was in 1970 (Bernhardt 2006). The FIRE sector has long existed in New York, is not growing much (and actually has been growing much less rapidly than the sector has elsewhere in the United States (see Center for an Urban Future 2003; see also Adler 2002 for a discussion of the changing demand for labor in the city's economy), and has significant local components (retail banks, insuring properties and autos in the city, etc.). Thus, rather than in the high-end producer services sectors, the growth in the city's employment has largely come from health care, personal services, and other services. The service sector, which is overwhelmingly localized services that are place-bound, has grown from 19 percent of the city's employment to more than 46 percent from 1960 to 2000 (Center for an Urban Future 2003). When these data are understood together, we are faced with a situation in which globalization has made employment the city much more localized, and much less exposed to global competition.

I do not want to make too much of this, because the city's economy is clearly globalized in myriad ways that extend beyond the composition of its employment. But given the central role that employment plays in people's lives, and the importance of exploitation (in the literal Marxist sense of the word) in the oppression of people—even if we accept that oppression is a multi-faceted social and structural construct (see Young 1990)—any discussion of economic justice and injustice must have employment at the center of its analysis. It is with this in mind that I now turn to a discussion of the growth of unregulated work in New York City, and the kinds of exploitation and economic injustice that is occurring in these jobs.

UNREGULATED WORK AND ECONOMIC INJUSTICE

Perhaps one of the most iconic expressions of the state of American cities for the workers in them is the growth of day-labor corners. These corners, which are not new *per se* (see Lionel Rogosin's 1956 film *On the Bowery*), have grown in their number, size, and significance in the last 20 years (Theodore 2003; Valenzuela *et al.* 2006). But day-labor corners are just the proverbial tip of the iceberg, as growing numbers of workers in American cities are working in what has variously been called, "precarious employment," "contingent employment," or "unregulated work" (Bernhardt *et al.* 2007).

These are jobs that, I would argue, are some of the most extreme expressions of economic injustice that exist in American cities. It therefore becomes imperative that we understand the jobs, industries, workers, and the causal forces behind growth of unregulated work. Over the course of several years of qualitative research we (myself, and researchers from the Brennan Center for Justice at NYU Law School) conducted almost 400 interviews about unregulated work with workers, employers, unions, government agencies, community organizations, legal service providers, and business/trade associations. In these jobs we have found a variety of different kinds of hyper-exploitative working conditions taking place in 14 different industries. These range from: domestic work (nannies, housekeeping), to industrial dry cleaners and laundries, to small-scale residential construction, to restaurants (see Table 8.1). The types of abuses of workers vary from industry to industry, but includes: not paying workers even the minimum wage (not to mention a living wage); not paying workers overtime; not paying workers at all; stealing workers' tips; unsafe, often deadly, working conditions; not providing workers' compensation when workers get sick or hurt on the job; and firing workers for engaging in any form of collective action (among many other kinds of abuses of workers).

As would be expected given the general discussion above on the composition of the city's employment, of the 14 industries we have identified, only one, garment manufacturing, is exposed to competition from outside

Table 8.1 Industries and occupations with workplace violations in New York City

Industry	*Industry segments with violations*	*Occupations most affected*
Groceries and Supermarkets	Green grocery stores, bodegas, delis, gourmet grocers, health food stores, non-union supermarkets	Cashiers, stock clerks, deli counter workers, food preparers, delivery workers, janitors, baggers, produce washers/watchers, and flower-arrangers
Retail (other than food)	Discount and convenience stores; ethnic retail; and to a lesser extent, non-union drug stores and retail chains	Cashiers, stock clerks, security guards, delivery workers, and workers in retailer-owned warehouses
Restaurants	All industry segments, especially high-end "white table cloth" restaurants and independent family-style and ethnic restaurants	Dishwashers, delivery persons, food prep, line cooks, porters, bussers, runners, bathroom attendants, barbacks, cashiers, counter persons, and coat checkers (and in some restaurants, waiters and waitresses and hosts)
Building maintenance and security	Non-union contractors providing services to small residential buildings and commercial clients; small residential and commercial buildings that hire workers directly	Security guards, janitors, supers, porters, handymen, and doormen
Domestic work	Individual families and diplomats	Nannies, housekeepers, housecleaners, elder companions, with many jobs combining duties from each
Child care	Publicly-subsidized home-based child care	"Legally-Exempt" and "Registered Family" child care workers

Home health care	Violations are common in the "gray market" where workers are employed directly by clients; some violations are also present for workers employed by home health care agencies	Home care workers
Construction	Small and medium private residential construction projects; small and medium public agency construction and renovation projects	Laborers, carpenters, and other construction trades
Manufacturing	Non-union food and apparel manufacturing	Sewing operators, machine operators, floor workers, pressers, hangers, packers, cutters, porters, and helpers/assistants
Laundry and dry cleaning	Non-union industrial laundries, dry cleaning plants, retail dry cleaners, and coin-op laundries	Folders, sorters, pressers, drivers, customer service workers, cleaners/spotters, tailors, and markers/baggers
Taxis	Yellow cabs, livery cabs, and dollar vans	Drivers
Auto services	Violations are common in car washes, but are also reported in informal parking lots, garages, and auto repair shops	Car wash workers and, to a lesser degree, parking attendants and auto body and repair workers
Personal services	Violations are common in nail salons, but are also reported in hair braiding shops, low-price spas hiring unlicensed massage therapists, and some beauty salons	Nail technicians, hair braiders and massage therapists, as well as other jobs in beauty salons such as attendants, janitors, and shampooers

Source: Bernhardt *et al.*, 2007.

of New York. And that industry has been declining rapidly in the city in the last 20 years, and looks likely to continue to decline in size (Fiscal Policy Institute 2003). All of the other sectors are localized services that are produced and consumed in place. These are jobs that are simply not leaving the city to go elsewhere. To be sure, many of these industries have emerged or grown because of the demand for them generated by the wealthy professionals who are employed in the producer services that lie at the organizing heart of the global economy. Saskia Sassen's well-known argument (Sassen 1991: Part III) of the cash-rich, but time-poor professional class driving demand for services—mostly reproductive services, that had formerly been provided within the household (usually by women)—in global cities certainly has echoes with many of the industries we have found to be most exploitative. And thus, the industries of domestic work, food preparation and delivery, etc., can rightly be explained through the lens of globalization, but only indirectly and, I would argue, peripherally (would it matter if the cash-rich and time-poor who contract out the services and goods of social reproduction were employed in industries serving a domestic or regional or local market, rather than a global one?).

And this explanation is only a small part of the industries discussed here. Some of the industries have gone through industrial restructuring on their own, largely in place—either literally *in situ* economic restructuring, or locally. The taxi industry, for instance, is one that has been restructured since the late 1970s, in ways that shift all of the burden and risk onto the workers (that is, drivers), while maintaining the control and profit by garages and medallion leasers or brokers (see Mathew 2005 for an excellent discussion of industrial restructuring in the taxi industry in New York). Drivers often will spend 12-hour shifts, working seven days a week for take home pay that averages $400–500 per week. Similarly, the supermarket industry has been restructured. Unionized supermarkets once dominated the industry, but are losing ground to a slew of non-union competitors—including gourmet grocers, health-food stores, greengrocers, big box stores, and drug stores that increasingly sell food items. The result has been a slow but steady deterioration in wages and benefits over time. In both industries, de-unionization coupled with changes in industrial structure have directly degraded the conditions of work, and led to working conditions that are hyper-exploitative.

Another way localized economic restructuring is leading to economic injustice is through contracting out of services. In some cases, this can be an occupation within a larger industry (janitors in supermarkets, for instance, who often get locked in supermarkets overnight by the stores, do not work for the supermarkets, but are employed by separate cleaning companies). Often, this can be the entire industry that is reshaped as a result of contracting out of services. For instance, industrial laundries have grown dramatically in the last 20 years, as hospitals and hotels that used to do their own linens, increasingly contract those services out to laundry facilities in the south Bronx or central Brooklyn. In these jobs workers are exposed to an astonishing variety of health and safety issues. Cleaning fluid (perchloroethylene) is carcinogenic, but perhaps worse than "perc" are the biological hazards that come with working in industrial laundries. Workers describe maggots coming out of laundry bags and, in the case of hospital linens, finding blood, needles, body parts, bits of fingers, etc.

In two of the industries where unregulated work is commonplace, the deregulation is the direct result of public policies driving the industries' growth. Subsidized childcare has grown dramatically since the 1996 welfare reform bill was passed, which drove many single mothers into the paid labor force. Most of this childcare is being provided by people whom the government calls "legally exempt" service providers. These workers, who are defined by the government as independent contractors, and thereby exempt from most basic labor laws in the United States, take care of two children in their own home. The government pays them $105 a week for full-time childcare, totaling $210 per week for 40 or more hours of work. With regard to home health care, the public-sector drivers are a bit more indirect, but just as vital. While most long-term care for the elderly is still done in nursing homes and other institutional settings, a rapidly increasing share is taking place in people's own homes. Medicare, the social health insurance for the elderly, pays for increasingly short hospital stays, and thereby more care is being provided outside of institutions, and in people's homes. Medicaid, the social health insurance for low-income people, then pays the long-term home health care costs. But Medicaid payments get contracted and sub-contracted out to private agencies several times on the way to paying the care givers (that is, the workers), and thus workers' wages are poor (SEIU 1199 2003). This is "actually existing neo-liberalism" (Brenner and Theodore 2002) as the state simultaneously cuts back on services, while contracting out services to the private (for-profit or not-for-profit) sector.

There is not the space here to fully explain the processes at work in all 14 industries, but a couple of things need to be made explicit. First, exploitation, and economic injustice, in these most exploitative and unjust jobs, has almost nothing to do with competition between people in New York and other cities. These are simply place-based employer–employee or capital–labor relations. This is about capitalists operating without the constraints of the state, or labor unions, or other organizations or interventions that limit the capacity of capital to exploit labor. Second, while the discussion here has been about New York City, parallel research currently being conducted in Chicago and Los Angeles is finding remarkably similar results (see DeFilippis *et al.* forthcoming, for a discussion of Chicago and New York City). The particular composition of the industries is a bit different, and both Los Angeles and Chicago have maintained larger manufacturing sectors than New York City (and Chicago has more warehousing, while Los Angeles has more short-haul trucking associated with the port), but the bulk of the unregulated work in all three cities is found in localized and personalized services. This is, simply put, not a problem in New York City; this is a problem in American cities.

Finally, it should also be noted that addressing, and mitigating (at a minimum) these exploitations would not seem to have much impact on the sectors exposed to external competition that are presently in the city. CBS is not going to make its locational decisions based on how much it costs to deliver Chinese food at night; and Wall Street executives are not going to push their firms to move elsewhere because the condo they bought on the Upper East Side for $2.3 million is going to cost an extra $50,000 to rehab. This assertion is borne out by what employers themselves are saying. After conducting a comprehensive survey of business owners in a host of sectors of the economy, the Bloomberg administration stated that the biggest concerns the employers had were around the issues of space and the cost of land. In fact, the Bloomberg administration's white paper on the issue bluntly stated: "payroll costs are the largest cost for the businesses surveyed, but employers consider the local workforce a valuable asset and did not cite payroll costs as a concern" (City of New York 2005: 13).

Rather than relocation, the bigger competitive issue for these sectors of the economy is with households providing (again) those services for themselves. As Esping-Anderson usefully argues:

> It is too often assumed that, since they are largely protected from international competition, services can provide a safe haven of employment—service workers do not compete with, say, Malaysians . . . True, in most economies the lion's share of services are sheltered from global competition. However, they face an even more ferocious competitor, namely family self-servicing.
>
> (1999: 115)

His argument, which I find convincing, is if prices for such services increase, then the difference between the wages of those buying the services and those providing the services will decrease. Thus, income inequality (what he calls, "price differentials") will decrease, and one of the driving forces in the growth of these services will be mitigated or significantly cut (and see Milkman *et al.* 1998, for an excellent discussion of the relationship between income inequality and domestic work in the United States). Jobs will thus disappear and households will re-internalize the provision of those services by (usually) women, expanding their "second shift" at home. But in what sense would that be more just as an outcome? Why should the goal be that people work full-time jobs for a salary and then work full-time jobs not for a salary (especially since the gendered division of labor in that second shift remains so pronounced)? Or that a very significant component of the workforce be made redundant? Thus the solution seems to be to either provide such services directly through the public sector, and thereby (potentially) increase the wages of the workers providing those services, or subsidize those services at the household level in order to increase wages without eliminating jobs.

What this whole discussion points to is a need to reframe the entire understanding of the relationships between competition, place, and urban economic justice. Competition, and its relationship to economic justice, needs to be recast as competition between people within places or cities—competition for both control over the means of production and the wealth that result from production.

The economic functions that occur in cities have not diminished in importance as a result of the globalization of the economy. And the central role of *the local* in what constitutes cities' economies has been re-articulated and transformed, rather than diminished in the contemporary political economy. The organization of work and labor markets is at the heart of the economic and political importance of cities—perhaps no less than at any time in the history of capitalism. As Peck has argued repeatedly (see, for instance, Peck 1996; Peck and Theodore 2007), labor markets are locally organized, produced, and regulated—and therefore labor markets need to be understood as localized processes.

The fundamental problem with the work of people like Castells (1977, 1983) and the whole language and framework of "collective consumption" is that it erases from cities—explicitly and consciously—the labor relations that exist within them. It is as if retail establishments were all self-serve, our residential buildings constructed and maintained themselves, our elderly in long-term care managed on their own, and our children looked out for themselves during the day when we worked at "real jobs." I am being a bit glib here, for sure, but the point is both intellectually and politically a vital one in understanding the potential for urban social movements to realize urban economic justice. Intellectually, what Castells and others have done is to confuse the outcome of production with the relations of production. Workers are workers in capitalism based on their class position, regardless of the material content of their work. As Marx (1963, vol. 1: 154) put it (in response to Adam Smith's false dichotomy between productive and unproductive labor):

> The determinate material form of the labor, and therefore of its product, in itself has nothing to do with the distinction between productive and unproductive labor. For example, the cooks and waiters in a public hotel are productive laborers, in so far as their labor is transformed into capital for the proprietor of the hotel.

It is only if we erase the class relations that constitute the service industries that we can say that cities are not the scale of production in the economy. And it will only be through recognizing the various ways in which the service industry is simultaneously place-bound and often characterized by exploitative workplace relations that we can begin to challenge the injustices that define urban economies.

EFFORTS FOR ECONOMIC JUSTICE IN AMERICAN CITIES

Fortunately, it is precisely this kind of understanding of cities as places of localized employment that has yielded some of the most exciting and dynamic urban social movements of the last decade. This chapter will end by briefly discussing four of them.

First, the living wage movement has emerged in the years since the first living wage ordinance was passed in Baltimore in 1994 to be a significant urban social movement. There are now more than 120 living wage ordinances in the United States, most at the municipal level (Luce 2007). These always apply to municipal government workers, with some ordinances applying to workers that are providing contracted government services, and even fewer ordinances applying to all workers in a municipality.

Second, there has been remarkable growth in the number of immigrant worker centers in North American cities in the last two decades. These centers are an important innovation in the field of community organizing, focusing issues of labor outside of the workplace. Gordon (2005: 280) describes these centers as seeking ". . . to build the collective power of their largely immigrant members and to raise wages and improve working conditions in the bottom-of-the-ladder jobs where they labor." It is estimated that there are approximately 140 centers of this type across the United States (Fine 2006). These are mainly in immigrant communities and work with those closest to the bottom of the labor market. They basically perform three, closely related, roles: they provide services to their members (often legal services; for instance trying to get unpaid wages); they do research and advocacy for their members (this is particularly true in the context of the local anti-immigrant ordinances); and they organize their members to get them more politically empowered.

Third, there are industry-specific organizing efforts. Sometimes this comes in the form of formalized industry-specific non-union worker organizations, and sometimes it comes in more fleeting organizational forms. In New York City there are several industry-specific organizing entities that work to improve the conditions of work in their industries. These include: 1) The Restaurant Opportunities Center of New York (or ROC-NY), a group which emerged from workers who had been employed at Windows on the World at the top of the World Trade Center and has since been organizing workers in the industry to improve working conditions industry-wide. 2) Domestic Workers United, an organization which emerged from Filipina domestic workers in 1999, but has since become a multi-ethnic and multi-racial organization, with political victories to its credit, ongoing aggressive organizing campaigns (for instance, to have a New York Statewide "Domestic Workers Bill of Rights"), and service delivery as well. 3) The Taxi Workers' Alliance (TWA) emerged in the latter part of the 1990s, in an industry in which a corrupt, formal union had long since abdicated its responsibilities to its workers. In the years since then, the TWA has waged a set of successful strikes and campaigns, including getting fare hikes (and having drivers keep a larger share of those hikes); fighting for money after 9/11 to compensate drivers for lost wages; and against increasing control over drivers' lives by the city and the garages.

Finally, worker cooperative businesses have demonstrated themselves to be able to not simply give workers more control over their jobs (through their ownership of their companies), but also improve working conditions for workers, and this is particularly true in service industry jobs, where conditions are often the most exploitative. Co-ops like Cooperative Home Care Associates in the Bronx, or UNITY Housecleaners in Hempstead, Long Island are demonstrating the potential for worker ownership in these industries.

Together, these four examples are probably where we can see the most productive calls for urban economic justice emerging, by organizations that understand the urban not just as a space of consumption, but as a space that is a mix of productive and reproductive labor, and of production and consumption.

CONCLUSION

Finally, I want to end this chapter by bringing the state back in to this discussion. These movements for economic justice have an interesting relationship to, and understanding of, the state in the urban political economy. The state, in these movements, is not viewed as the source of redistribution. Instead, the state is viewed as the regulatory framework that governs our lives—but the redistribution sought is from capital to labor. The state facilitates this transfer, but does not provide it (as it does through public assistance programs and other mechanisms).

There are clearly echoes here with many of the state theories of the 1970s and early 1980s which addressed how states came to absorb conflict that emerged in other sectors (that is, between capital and labor), and by virtue of their absorption of those conflicts the state became the arena of class conflict. But by absorbing and displacing that conflict, and shifting political attention to arenas where it should not be, the state enabled the reproduction of the unjust capitalist political economy (see Barrow 1993 for a useful summary of these debates; see, in particular, Hirsch 1978 for a good example of this argument; and see Piven and Cloward 1971 for a closely related set of arguments about the uses of the welfare state in capitalism). Put another way, and moving the discussion to the scale of the city, urban Keynesianism transformed cities and their politics in important ways. As Harvey (1985: 37–38) put it:

> The Keynesian city was shaped as a consumption artifact and its social, economic, and political life organized around the theme of state-backed, debt-financed consumption. The focus of urban politics shifted away from alliances of classes confronting class issues towards more diffuse coalitions of interests around themes of consumption, distribution, and the production and control of space . . .

But the Keynesian city as a social process is now behind us, and the state has now removed itself from much of its social redistribution role. And while there have been huge problems that have resulted from that withdrawal, it has also opened the space back up for social movements that are focusing their attention and efforts at capital. Class has re-emerged in the forefront of movements for social justice in cities.

But the withdrawal of the state from its redistributive role has been mirrored by its shirking from its regulatory role as well. And thus all of these efforts have emerged, or are emerging, in a context in which the institutional forms that should be limiting the capacity of capital to exploit labor have been unable or unwilling to perform this task—most notably the state, but also labor unions. The fight, therefore, is to rebuild the local regulatory capacity to protect workers in their struggles for economic justice in American cities. Thus one of the primary goals of planners interested in economic justice should be to find ways to support, facilitate, and enable those movements.

The road to economic justice is being paved by those who:

a) understand the nature of urban economies;
b) are using that understanding to challenge the injustices within those economies;
c) and are pushing for, and often getting, local political and economic victories as a result.

We, as planners and urbanists, have to find our voice and our role in these efforts.

NOTE

1 Just to be clear, this is not competition between *cities, per se*, since cities are not, strictly speaking, actors that can compete (see Marcuse 2005).

REFERENCES

Adler, M. (2002) "Why Did New York Workers Lose Ground in the 1990s?," *Regional Labor Review*, Fall: 31–35.

Barrow, C. (1993) *Critical Theories of the State: Marxist, Neo-Marxist, Post-Marxist*, Madison: University of Wisconsin Press.

Bernhardt, A. (2006) Presentation to the Re-Defining Economic Development in New York coalition, April 25, New York.

Bernhardt, A., McGrath, S. and DeFilippis, J. (2007) *Unregulated Work in the Global City: Employment and Labor Law Violations in New York City*, New York: Brennan Center for Justice at New York University Law School.

Brenner, N. and Theodore, N. (eds) (2002) *Spaces of Neoliberalism: Urban Restructuring in North America and Western Europe*, Oxford: Blackwell.

Castells, M. (1977) *The Urban Question*, Cambridge, MA: MIT Press.

—— (1983) *The City and the Grassroots*, Berkeley: University of California Press.

Center for an Urban Future (2003) *Engine Failure*, New York: Center for an Urban Future.

City of New York (2005) *New York City Industrial Policy: Protecting and Growing New York City's Industrial Job Base*, New York: City of New York.

DeFilippis, J., Martin, N., Bernhardt, A., and McGrath, S. (forthcoming) "On the Characteristics and Organization of Unregulated Work in American Cities," *Urban Geography*.

Esping-Anderson, G. (1999) *Social foundations of Postindustrial Economies*, Oxford: Oxford University Press.

Fine, J. (2006) *Worker Centers: Organizing Communities at the Edge of the Dream*, Ithaca: Cornell University Press.

Fiscal Policy Institute (2003) *NYC's Garment Industry: A New Look?*, New York: Fiscal Policy Institute.

Fitch, R. (1993) *The Assassination of New York*, New York: Verso.

Friedman, T. (1999) *The Lexus and the Olive Tree*, New York: Farmer, Strauss and Giroux.

Gordon, J. (2005) *Suburban Sweatshops: The Fight for Immigrant Rights*, Cambridge, MA: Harvard University Press.

Harvey, D. (1985) *The Urban Experience*, Baltimore: Johns Hopkins University Press.

—— (1996) *Justice, Nature, and the Geography of Difference*, Oxford: Blackwell.

Hirsch, J. (1978) "The State Apparatus and Social Reproduction: Elements of a Theory of the Bourgeois State," in Holloway, J. and Picciotto, S. (eds) *State and Capital: A Marxist Debate*, Austin: University of Texas Press.

Luce, S. (2007) "The U.S. Living Wage Movement: Building Coalitions from the Local Level in the Global Economy," in Turner, L. and Cornfield, D. (eds) *Labor in the New Urban Battlegrounds: Local Solidarity in a Global Economy*, Ithaca: Cornell University Press.

Marcuse, P. (2005) "'The City' as Perverse Metaphor," *Cities*, 9(2): 247–254.

Marx, K. (1963 and 1863) *Theories of Surplus Value*, Burns (trans.), vols. 1–3, E., Moscow: Foreign Languages Publishing House.

Mathew, B. (2005) *Taxi! Cabs and Capitalism in New York City*, New York: The New Press.

Milkman, R., Reese E., and Roth, B. (1998) "The Macrosociology of Paid Domestic Labor," *Work and Occupations*, 25(4): 483–510.

Peck, J. (1996) *Work Place: The Social Regulation of Labor Markets*, New York: Guilford Press.

Peck, J. and Theodore, N. (2007) "Flexible Recession: The Temporary Staffing Industry and Mediated Work in the United States," *Cambridge Journal of Economics*, 31(2): 171–192.

Peterson, P. (1981) *City Limits*, Chicago: University of Chicago Press.

Piven, F. (1995) "Is it Global Economics or Neo-Laissez–Faire?," *New Left Review*, 213: 107–114.

Piven, F. and Cloward, R. (1971) *Regulating the Poor*, New York: Pantheon Books.

Pratt Center for Community and Environmental Development (2001) *Making It in New York: The Manufacturing Zoning and Land Use Initiative*, New York: Municipal Arts Society.

Sassen, S. (1991) *The Global City: New York, London, Tokyo*, Princeton, New Jersey: Princeton University Press.

SEIU Local 1199 (2003) *The Plight of New York's Home Health Aides*, New York: SEIU Local 1199.

Theodore, N. (2003) "Political Economies of Day Labour: Regulation and Restructuring of Chicago's Contingent Labor Markets," *Urban Studies*, 40, 1811–1828.

Tiebout, C. (1956) "A Pure Theory of Local Expenditures," *The Journal of Political Economy*, 64(5): 416–424.

Valenzuela, A., Theodore, N. Meléndez, E., and Gonzalez, A.L. (2006) "On the Corner: Day Labor in the United States," January, available at: http://www.uic.edu/cuppa/uicued/Publications/RECENT/onthecorner.pdf (accessed August 15, 2008).

Wilson, D. (2007) *Cities and Race: America's New Black Ghetto*, New York: Routledge.

Young, I. (1990) *Justice and the Politics of Difference*, Princeton: Princeton University Press.

Part III

How do we realize Just Cities?

From debate to action

9 Keeping counterpublics alive in planning[1]

Laura Wolf-Powers

INTRODUCTION: JUSTICE AND THE PUBLIC SECTOR

The editors of this volume assert that a Just City—the "path between the universal and the particular" that Susan Fainstein discusses in Chapter 1—can only be developed by participants in practice (see Introduction). More than any ideological stance or unifying theoretical rubric, the goal of justice in practice defines what I will call here the activist or progressive wing of the city planning profession. A sense of the latent possibility that state institutions, responsive to the disadvantaged and vulnerable in addition to the powerful and well-resourced, might achieve something that resembles a fair distribution of opportunities and pleasures within urban places[2] draws idealistic young people into planning in the United States today as it did in the Progressive Era and in the turbulent 1960s and 1970s. Some proportion of planning school graduates—perhaps not a majority, but some—will enter professional life with the aim of justice in mind.

Many academic planners define and explore justice in theory. But part of what makes the profession of planning appeal to students is the practical and place-based nature of its search for alternatives to the social given. Of chief relevance to planning (in its conventional as well as its more radical forms) is what is happening in places and to people in places; it engages the everyday public sphere. Why did the little hardware store down the street go out of business? How can we make it easier to commute by bicycle? What will the quality of place in this neighborhood be like once the proposed redevelopment is enacted, and who, exactly, will be enjoying the quality of the redeveloped place? Questions of justice are latent in all of these formulations, but the problems they raise are also quite practical. John Friedmann (1987) recognizes this relationship between practical and theoretical definitions of justice when he characterizes planning as an attempt "to link scientific and technical knowledge to actions in the public domain" (p. 38) and when he highlights "the problem of how to make technical knowledge in planning effective in informing public actions" (p. 36).[3]

Friedmann's clear view is that attempts by planners to realize justice—or as he terms it, to make "social rationality" prevail over "market rationality"

—require the state. He notes that when public action involves the contravention of market principles in the name of social interests, conflict ensues in the state domain that is often resolved in favor of market actors. But in the context of wider political mobilization, he argues, planners, including planners in government, can act in and on the public domain in ways that fulfill the goal of creating more just places. This notion also lies at the core of the "equity planning" and "progressive city" literature, which highlights the accomplishments of public sector leaders who strive, within constrained arenas of power, to allocate resources and make decisions as though the interests and rights of poor and working-class people mattered (see Clavel 1986; Krumholz and Forester 1990; Clavel and Wiewel 1991; Krumholz and Clavel 1994).

In the two decades since Friedmann published *Planning in the Public Domain*, the state's limited range of motion on behalf of broadly defined social interests (redistribution of income, environmental protection, a right to housing) has only been further curtailed, particularly in the United States. Work in geography, sociology and political science documents and laments the rise and hegemony of neo-liberal ideology as a guide for public policy and state action. The achievements of equity planners and progressive mayors are overshadowed by the work of the global vectors of capital accumulation,[4] as well as by more local institutions—development authorities, for example—with the power to bypass planners and subvert public process. Yet every year, cadres of students who see city planning as a route to lessening urban injustice and misery enroll in, and graduate from, master's degree programs. Some of them go to work for "movement" organizations whose *raison d'être* is to shape public policy from outside, but most find places either in government agencies or in non-profit groups that rely on the state for survival. A central question for progressive planning, therefore, is that of how the state sphere, lying as it does at the conflict-ridden heart of the planning field, might be shaped (from without and within) in the interests of justice.

The details of the Bronx Terminal Market case elucidated by Fainstein in Chapter 1 of this volume suggest the enormity and complexity of this task. In the Bronx, city government actors, in the name of public-serving redevelopment, denied a group of wholesale merchants and their four hundred employees both their means of livelihood and recognition as legitimate stakeholders in land use determination. This and other examples lead Fainstein to concur with Friedmann that social mobilization has a critical role to play in justice-oriented planners' ability to exercise influence on the state: "The movement toward a normative vision of the city requires the development of counter-institutions capable of reframing issues in broad terms and of mobilizing organizational and financial resources to fight for their aims" (p. 35).

As Fainstein recognizes, issue-framing capabilities within the dominant public sphere are limited, as are resources for marginal groups. The Bronx Terminal Market merchants were unable to prevail in court or convince the City Planning Commission or City Council to take them seriously, both because they lacked influence and because they were unable to counter "the logic that the new mall represented necessary modernization" (p. 23). Yet in invoking both the social movements of the 1960s and contemporary living wage movements (p. 34), Fainstein reveals faith in the possibility that urban social movements might still be capable of influencing public policy, implying a need for progressive planners to strategize both about mobilizing resources for such groups and about reframing urban issues in public discourse.

It is the question of issue-framing that engages me here. The remainder of this chapter is dedicated to exploring how "marginal publics" or "counterpublics," in the parlance of communications theory, do rhetorical work that advantages the interests of marginalized groups with respect to the dominant public sphere and the state. I argue that city planners have a part to play in the formation and support of counterpublic discourses which, in dialogue with the state and from within the state, influence public and social policy relevant to cities and their inhabitants.

DEFINING COUNTERPUBLICS

The use of the terms "public" and "counterpublic" varies among scholars. The concepts originally derived from Habermas' (1989) account of how a "*bourgeois* public sphere" arose alongside modern constitutional government in eighteenth- and nineteenth-century Europe and North America as a milieu in which people debated the activities of the state (see Asen and Brouwer 2001; Squires 2002).[5] Alternative conceptions of the public sphere were soon proposed by theorists who criticized Habermas for idealizing a construct that excluded women, most persons of color, and the working class (see, e.g., Fraser 1992). Gradually a notion of multiple and competing public spheres emerged. Central to this is the idea of a dominant public sphere that expresses the social and political hegemony of dominant class social interests, surrounded and interpenetrated by subordinate or "subaltern" public spheres in which marginal groups consolidate oppositional identities and circulate alternative interpretations (both positive and normative) of the world. A counterpublic sphere, as a space of conversation, performance and argument rather than concerted action, is not the cradle of a social movement. However, in many scholarly accounts, counterpublics provide crucial vocabularies and framing devices which social movement actors carry into the world in order to articulate their interests and needs before dominant publics and the state.

A counterpublic discourse documented by communications scholar Devorah Heitner in her work on Black public affairs television is embodied in the show *Inside Bedford Stuyvesant*, produced in New York City from 1968 to 1971 (Heitner 2007). The program, a news magazine about the historically African-American neighborhood (with a population of 400,000 that encompassed poor, working-class and middle-class households), was brought into being through a collaboration between the local television station WNEW and newly formed community development group, the Bedford-Stuyvesant Restoration Corporation (BSRC).[6] *Inside Bedford Stuyvesant* was created in a political milieu that had been heavily influenced by Daniel Patrick Moynihan's Report *The Negro Family: The Case for National Action* as well as the *1968 Kerner Commission Report*. Contemporary government discourse on urban African-Americans centered on their pathology and that of the "ghetto" communities in which they lived. Hosted by BSRC Associate Director James Lowry and television personality Roxie Roker, the show countered these recurrent public themes by documenting Bedford-Stuyvesant itself (the episodes were filmed in the community, often outdoors) and amplifying the voices of concerned, active African-Americans of diverse philosophies and priorities who lived in and cared about the neighborhood. The program served as a venue for commentary and debate on local and national politics and culture by figures as diverse as Harry Belafonte, Sonny Carson, and Amiri Baraka as well as local "experts" whose recognition derived simply from the fact that they lived in the neighborhood and had a stake in its future.

Heitner argues that *Inside Bedford Stuyvesant* portrayed the neighborhood's residents "as citizens for whom political ideas for transforming space and community are omnipresent and debated" (p. 90), as against the image of apathetic ghetto dwellers projected in the dominant public sphere. In no sense was the program affiliated with a particular social movement or political stance; in fact, it painstakingly presented a variety of perspectives and consciously made itself palatable to a general audience through such strategies as choosing "ambassadorial" hosts with "middle-class linguistic styles and appearance" (p. 61). However, people who contributed to and watched the program and identified with its agenda participated in social movements (for instance the Black Arts Movement, the welfare rights movement, and the community development movement) that helped to reframe mainstream attitudes and policies affecting the city's African-American communities. As they were nourished and inspired by alternative portrayals of their own community, activists in Bed-Stuy and their counterparts in other neighborhoods found that the show helped to increased receptivity in the broader public arena to their policy goals.[7] During this time period, community development activists (including staff and board members of the Bedford-Stuyvesant Restoration Corporation) helped to shift New York City housing practice from one of condemning blocks and demolishing abandoned buildings in "slum" neighborhoods to one of supporting local non-profit organizations in rehabilitating housing and developing vacant property. Also during this time, activists successfully campaigned to redraw Brooklyn's congressional district boundaries, a move that resulted in the election of the nation's first Black Congresswoman, Shirley Chisholm, to represent the Bedford-Stuyvesant neighborhood.

A PLANNING COUNTERPUBLIC

Another counterpublic discourse, also drawn from the Brooklyn of that time period but more directly connected to the city planning profession itself, can be found in *STREET* magazine. *STREET*, with a circulation of 5,000 throughout New York City but mainly in Brooklyn, was published roughly quarterly by the Pratt Institute Center for Community and Environmental Development from 1971 to 1975 and directed to an audience of both professional planners and activists (see Wolf-Powers 2008).[8] The publication was conceived and created in a time when many planners were countering mainstream practice, drawing on the alternative visions of society that were animating the civil rights and economic justice movements. Inspired by critical scholarship documenting urban renewal's impact on the poor (Gans 1959; Marris 1962) and by the revelatory literature on inner-city poverty that had influenced the Kennedy Administration (see Lemann 1991; Halpern 1995), they espoused a vision of planning that encompassed the entire society, not just the physical city (Pynoos *et al.* 1980, 2002; Hartman (2002); Clavel 1986; Hoffman 1989). Many within this new wing of the profession, like the so-called "social progressives" who had been prominent in the city planning movement in the early twentieth century (see Peterson 2003), saw planners as a movement group whose members might seek to uproot poverty and challenge inequitable social arrangements even as they worked within existing legal and institutional contexts (Davidoff 1965; Clavel 1986; Krumholz and Clavel 1994). For example, in the context of a burgeoning national environmental movement, *STREET* published reports and bulletins on national and state legislation and summaries of research findings on pollution, but differed from much contemporary environmental discourse in that it presented the environment as an urban concern and drew connections between race, class, and neighborhood environmental quality (see Figure 9.1). It also emphasized neighborhoods and households as sites for the expression of social values, reporting homegrown efforts to encourage recycling, alternative transportation, urban agriculture, and the consumption of local food.

At the same time, during a period of severe disinvestment and distress in New York's working-class and low-income communities, *STREET* offered an upbeat and defiant counterpoint to official narratives of decline, chronicling both formal neighborhood-based planning efforts and informal citizen-initiated activity. Features depicting Brooklyn residents going to church, attending block parties, building sweat equity housing, maintaining small businesses and caring for their families and front yards link *STREET* to the more widely consumed *Inside Bedford-Stuyvesant* as a force of counter-representation (see Figure 9.2). Features criticizing Federal Housing Administration policy and New York City's allocation of Community Development Block Grant funds and applauding state representatives who had introduced anti-redlining legislation positioned *STREET* as a forum for serious discussion of public policy.

Because *STREET* was connected with a technical assistance planning organization, it fed and documented advocacy work more directly than *Inside Bedford Stuyvesant*. For example, articles in Issues IX and XIV of *STREET* reported on a lawsuit that a group of block associations had filed seeking a judgment against the U.S. Department of Housing and Urban Development requiring that it study the potential harmful environmental impact of

Figure 9.1 Cover of *STREET* Magazine, Issue IV. Illustrated by Uffe Surland

its "delivered vacant" policy. This was a policy requiring that federally insured buildings whose owners had defaulted on their mortgages be vacated of tenants and stripped of value in order for lenders to collect government-issued insurance. The "delivered vacant" policy, litigants argued, not only prompted the eviction of indigent tenants; the improper sealing of the vacant buildings encouraged vandalism and destabilized conditions for families who remained in the neighborhoods. This and other legal action ultimately suc-

Figure 9.2 Photographs accompanying an article about a Brooklyn block party. *STREET* magazine, Issue VII

ceeded in forcing HUD to change its protocols (see also Kramer 1974). The counterpublic stance of *STREET*, which had consistently questioned the logic and the morality of federal housing policy, in this case nourished local residents' effort to confront HUD and galvanized the lawyers who filed motions on their behalf.

STREET both reflected the emergence of an "activist turn" in city planning and provided the planners in its audience with images of themselves and a picture of the world that helped solidify their identities as oppositional actors. Because *STREET*'s public included many key actors in New York City's neighborhood environmental and housing movements, the magazine helped reframe neighborhood development questions in ways that arguably had an impact on public policies. For example, over the course of the 1970s, mainstream city government institutions gradually embraced the role of community-based organizations in developing and managing low- and moderate-income housing, making CDC and CBO involvement virtually an article of faith in the system of social housing production (Goetz 1996; Rosen and Dienstfrey 1999).[9] The magazine thus made a small contribution to a counterpublic discursive arena in the field of urban redevelopment. But the magazine's more effective "counterpublicity" related not to dominant understandings of urban development but to the dominant understanding of the goals of city planning. Along with others in the progressive wing of the profession during this time period, *STREET*'s creators countered contemporary understandings of what constituted an adequate professional response to disinvestment from central city neighborhoods, what planners could and should do with their professional skills, and who could legitimately plan.

LOOKING AHEAD

We are no longer in that era. The legacies of the alternative conception of city planning promulgated by progressive forces in the era of *Inside Bedford Stuyvesant* and *STREET* can be found in the continuing engagement of some planners with environmental justice, place-based community development and progressive regionalism; but in comparison to what existed in the 1960s and 1970s, activist planning looks anemic. Justice-motivated planners lack the infrastructure of federal funding for and endorsement of urban community development and anti-poverty initiatives that supported their project four decades ago.[10] In many cities (though certainly not all), the political-economic context of housing and planning has also changed; Bedford-Stuyvesant still struggles with poverty, but it is market-rate reinvestment and predatory lending rather than property abandonment and the urban renewal bulldozer that pose the greatest displacement threats to vulnerable residents. Finally, community-based planners, perhaps by necessity, now have a different relationship to the local state. In the early 1970s, when federal funding was more plentiful and both local government and finance capital appeared to have given up on central cities, activist planners thrived in a position of relative autonomy from local bureaucracies, forming renegade organizations that built and managed housing, created public spaces and mobilized neighborhood residents in local politics. Community development has now been assimilated more fully into the state's purview, and the tools embraced by those individuals who created and consumed *STREET* magazine are more difficult to put into play (see Stoecker 2003, 2004; Lander 2005, DeFilippis *et al.* 2006; Marwell 2007).

The federal support that funded the emergence of counterpublic voices in *Inside Bedford-Stuyvesant* and *STREET* (and that often financed the activists they informed) is absent from the current urban political economy. Nevertheless, city planning's traditional emphasis on the local, the concrete and the here-and-now can be a source of strength to activists attempting to assert the needs of disadvantaged groups in the context of the "market rationality" of conventional urban development. In contrast to the past, when staff of the Pratt Center pasted together issues of *STREET* by hand and the producers of *Inside Bedford-Stuyvesant* shouldered huge cameras, the variety and accessibility of the media advantages the issue-reframing narratives of counterpublic discourse. One example in contemporary New York City is the profusion of websites and blogs—such as Norman Oder's Atlantic Yards Report—on which people debate the impact of redevelopment projects on both quality of life and the distribution of wealth.[11]

Another strong suit of planning goes back to Friedmann's conviction that in the actually existing political economy of our society, justice requires the state. Counterpublic strategies in planning are by definition opposed to the patterns of thought and habits of mind that characterize much urban development and policy. Because of our role in the creation and transmission of action-oriented knowledge (Friedmann 1987; Forester 1988; Hoch 1994; Throgmorton 1996), and our proximity to the public domain, city planners are in a good position to "shake things up" both by offering expertise to social movement actors and by introducing alternative visions of the city into the dominant public discourse.

Planners, well acquainted with the analysis of political power in local government settings, are also cognizant of what is possible and achievable in the context of existing societal relations. Social criticism that takes public sector institutions to task for their loyalty to "market rationality" inspires important questioning and protest. But the critique is not enough. One recent example of incisive but ultimately damaging critique is Walter Thabit's (2003) compelling book *How East New York Became a Ghetto*, which tells of his experiences as a planner in eastern Brooklyn in the 1960s and 1970s. While Thabit bears painful witness to the calamity that befell the community and its residents during this time, he offers few clues to readers who want to understand how different actions, choices or institutional arrangements *in the public sector* might have led to different outcomes. The book simply excoriates government for its indifference and venality, providing ballast for the arguments of market fundamentalists for whom government failure is the only explanation for urban decline.

Progressive planners inside and outside of the public sector must consider how, in the messy reality that is a city, planners can nudge, prod or drag government institutions to create policy mechanisms through which resources and opportunity are more equitably shared. The "equity planning" literature of the late 1980s merits an update; perhaps its focus today would not be the previous accomplishments of enlightened leaders but prospective strategies to consider under current circumstances. Work on equitable development being undertaken by groups such as *PolicyLink* provides a way forward nationally; many cities and metropolitan areas also have "action research" agencies or "think and do tanks" concerned with promoting just planning and development in the political economies of particular places, Strategic Action for a Just Economy in Los Angeles and the Philadelphia Unemployment Project being just two examples.

Well before the counterpublic discourses discussed in this essay had taken hold in city planning, a cartoon by Saul Steinberg, picturing a man on a winged horse riding on the back of a slow-moving turtle, was reprinted by Professor George Raymond, the first Chairperson of the Department of City and Regional Planning at Pratt Institute, in Volume 1 Issue 4 of a publication called the *Pratt Planning Papers* (Department of City and Regional Planning, Pratt Institute 1963).[12] Raymond added the wry caption: "We continue our inquiry into the nature of the planning profession." One interpretation of our job as planning educators is that we cement our students to the turtle's back. I prefer to hope that we help them infuse the brain and legs of the pragmatic turtle with the horse's intelligence, imagination and will to flight.

NOTES

1 The author would like to acknowledge the valuable contribution of Justin Steil, who reviewed and edited this essay.

2 Other contributors to this volume concern themselves with an exact definition of the "Just City." This one, though surely assailable on many counts, stands as mine.
3 If this problem is not solved, argues Friedmann (1987), "planners will end up talking only to themselves and eventually will become irrelevant" (p. 36).
4 Krumholz's home city of Cleveland, for example, has experienced devastating employment loss and increases in poverty and inequality since his *Making Equity Planning Work* was published in 1990.
5 Habermas argues that the *bourgeois* public sphere subsequently declined amidst the contradictions of industrialization and the modern welfare state.
6 Perhaps significantly, the Bedford-Stuyvesant Restoration Corporation, which sponsored the show, was in part the outgrowth of a neighborhood planning effort undertaken in 1964 to evaluate a city-sponsored proposal for the Fulton Park Urban Renewal Area. The planning effort, whose sponsor, the Central Brooklyn Coordinating Council, drew the attention of the Ford Foundation and Senator Robert Kennedy, culminated in the founding of BSRC—the first federally-funded community development corporation—in 1967. The founding and history of the group have been documented by Johnson (2004) and by Ryan (2004). See also the Pratt Center for Community Development CDC Oral History Project (http://www.prattcenter.net/cdc-bsrc.php). Heitner notes: "The BSRC's mission to rehabilitate housing and stimulate economic development in Bedford-Stuyvesant did not prevent [*Inside Bedford Stuyvesant*] from hosting guests who were critical of some of the effects and methods of redevelopment there" (p. 86).
7 While the effect of a television program would ordinarily be a matter of inference, in this case a rare collection of letters from viewers housed in the archives of the Bedford-Stuyvesant Restoration Corporation and reviewed by Heitner (2007) provides evidence of the program's influence on a variety of viewers: members of the community, viewers of color from elsewhere, and whites.
8 The staff of the Pratt Institute Center for Community and Environmental Development printed 5,000 copies of each issue of *STREET*. Two thousand went by mail to neighborhood-based planners and community organization staff and leaders who had participated in the organization's seminars, trainings and conferences. Staff dropped off the remaining three thousand copies at the offices of local anti-poverty agencies, community organizations and community facilities like public libraries.
9 Pratt staff ceased to publish *STREET* in 1975 but worked with three neighborhood housing movement groups to launch the independent publication *City Limits* starting in 1976. *City Limits* chronicled and informed New York City's neighborhood housing and community development movement for the next 30 years.
10 The Bedford-Stuyvesant Restoration Corporation, which helped produce *Inside Bedford-Stuyvesant*, was funded by the U.S. Departments of Labor and Commerce. *STREET* magazine was funded by a grant from the Office of Environmental Education in the U.S. Department of Health, Education and Welfare.
11 This should not be exaggerated; media access continues to be skewed toward the wealthy, as demonstrated by the large number of sites and blogs that focus almost exclusively on development's aesthetic and congestion-related impact (as opposed to its effect on the poor). A notable exception to this trend is the website of the organization Good Jobs New York.
12 The *Pratt Planning Papers* were published by Pratt's planning department from 1963–67 and had no direct connection to *STREET* magazine.

REFERENCES

Asen, R. and Brouwer, D.C. (2001) "Introduction: Reconfigurations of the Public Sphere," in R. Asen and D.C. Brouwer (eds) *Counterpublics and the State*, Albany: State University of New York Press.

Clavel, P. (1986) *The Progressive City: Planning and Participation, 1969–1984*, New Brunswick, NJ: Rutgers University Press.

Clavel, P. and Wiewel, W. (eds.) (1991) *Harold Washington and the neighborhoods: Progressive city government in Chicago, 1983–1987*, New Brunswick, NJ: Rutgers University Press.

Davidoff, P. (1965) "Advocacy and pluralism in planning," *Journal of the American Institute of Planners*, 31(4): 334.

DeFilippis, J., Fisher, R., and Shragge, E. (2006) "Neither romance nor regulation: Re-evaluating community," *International Journal of Urban and Regional Research*, 30(3): 673–689.

Forester, J. (1988) *Planning in the Face of Power*, Berkeley: University of California Press.

Fraser, N. (1992) "Rethinking the public sphere: A contribution to the critique of actually existing democracy," in C. Calhoun (ed.) *Habermas and the Public Sphere*, Cambridge, MA: Massachusetts Institute of Technology Press.

Friedmann, J. (1987) *Planning in the Public Domain*, Princeton, NJ: Princeton University Press.

Gans, H. J. (1959) "The human implications of current redevelopment and relocation planning," *Journal of the American Institute of Planners*, 25 (1): 15–25.

Goetz, E. (1996) "The neighborhood housing movement," in W. Dennis Keating, Norman Krumholz, and Philip Star (eds) *Revitalizing Urban Neighborhoods*, Lawrence, KS: University Press of Kansas.

Habermas, J. (1989) *The Structural Transformation of the Public Sphere: An inquiry into a category of bourgeois society*, trans. Thomas Burger and Frederick Lawrence, Cambridge, MA: Massachusetts Institute of Technology Press.

Halpern, R. (1995) *Rebuilding the Inner-City: A history of neighborhood initiatives to address poverty in the United States*, New York: Columbia University Press.

Hartman, C. (2002) *City for Sale: The transformation of San Francisco*, Berkeley, CA: University of California Press.

Heitner, D. (2007) "Black power TV: A cultural history of black public affairs television 1968–1980," unpublished dissertation manuscript, Northwestern University.

Hoch, C. (1994) *What Planners Do: Power, Politics and Persuasion*, Chicago: Planners Press.

Hoffman, L. (1989) *The Politics of Knowledge: Activist movements in medicine and planning*, New York: State University of New York Press.

Johnson, K. (2004) "Community development organizations, participation and accountability: The Harlem Urban Development Corporation and the Bedford Stuyvesant Restoration Corporation," *Annals: Journal of the American Academy of Political and Social Science, Race and Community Development Issue*, 594(1): 187–189.

Kramer, D.J. (1974) "Protecting the urban environment from the federal government," *Urban Affairs Quarterly*, 9(3): 359–368.

Krumholz, N. and Clavel, P. (1994) *Reinventing Cities: Equity Planners Tell Their Stories*, Philadelphia: Temple University Press.

Krumholz, N. and Forester, J. (1990) *Making Equity Planning Work: Leadership in the public sector*, Philadelphia: Temple University Press.

Lander, B. (2005) "Community Development: Progressive and/or Pragmatic?," paper presented at City Legacies: A Symposium on Early Pratt Planning Papers and *STREET* magazine in New York, NY, Pratt Institute Manhattan Campus, October 14. See http://www.pratt.edu/newsite/xfer/citylegacies/index.php#schedule.

Lemann, N. (1991) *The Promised Land: The great black migration and how it changed America*, New York: A.A. Knopf.

Marris, P. (1962) "The social implications of urban redevelopment," *Journal of the American Institute of Planners*, 28(3): 15–25.

Marwell, N. (2007) *Bargaining for Brooklyn: Community Organizations in the Entrepreneurial City*, Chicago: University of Chicago Press.

Peterson, J.A. (2003) *The Birth of City Planning in the United States, 1840–1917*. Baltimore, MD: Johns Hopkins University Press.

Pratt Institute, Department of City and Regional Planning (1963) *Pratt Planning Papers* 1(4). Brooklyn, NY: available at: http://www.pratt.edu/newsite/xfer/citylegacies/PPP_volumes/VOL_1_NO_4.pdf (accessed July 12, 2008).

Pratt Institute Center for Community and Environmental Development (1971–75) *STREET* magazine, 1–15, Brooklyn, NY: available at: http://www.pratt.edu/newsite/xfer/citylegacies/downloads.php (accessed July 12, 2008).

Pynoos, J., Schafer, R., and Hartman, C. (1980, revised) *Housing Urban America*, Chicago: Aldine Publishing Co.

Rosen, K. and Dienstfrey, T. (1999) "Housing services in low-income neighborhoods," in Ronald F. Ferguson and William T. Dickens (eds) *Urban Problems and Community Development*, Washington, D.C.: Brookings Institution.

Ryan, W. P. (2004) "Bedford Stuyvesant and the prototype community development corporation," in M. Sviridoff (ed.) *Inventing Community Renewal: The Trials and Errors that Shaped the Modern Community Development Corporation*, New York: Community Development Research Center, New School for Social Research.

Squires, C. R. (2002) "Rethinking the black public sphere: an alternative vocabulary for multiple public spheres," *Communication Theory*, 12(4): 446–468.

Stoecker, R. (2003) "Understanding the development organizing dialectic," *Journal of Urban Affairs*, 25(4): 493–512.

—— (2004) "The mystery of the missing social capital and the ghost of social structure: why community development can't win," in Robert M. Silverman (ed.) *Community-Based Organizations: The Intersection of Social Capital and Local Context in Contemporary Urban Society*, Detroit: Wayne State University Press.

Thabit, W. (2003) *How East New York became a Ghetto*, New York: New York University Press.

Throgmorton, J. (1996) *Planning as Persuasive Storytelling the Rhetorical Construction of Chicago's Electric Future*, Chicago: University of Chicago Press.

Wolf-Powers, L. (2008) "Expanding planning's public sphere: *STREET* magazine, Activist Planning and Community Development in Brooklyn, NY 1971–75," *Journal of Planning Education and Research*, 28(2): 180–195.

10 Can the Just City be built from below?

Brownfields, planning, and power in the South Bronx

Justin Steil and James Connolly[1]

Writings on the Just City have advanced theoretical and philosophical justifications for a redefinition of planning priorities, but have not yet examined in detail issues of institutional structure. Governance of brownfield redevelopment in the United States provides one example where active institutional experimentation has in some cases been shaped by explicit articulations of justice. This chapter examines such an effort through a case study of a coalition of grassroots environmental justice organizations in New York City's South Bronx that have been working to reconfigure organizational relations in their neighborhoods for roughly a decade. While the analysis of the Bronx groups' experiences cannot be fully developed here, this chapter focuses on the environmental justice organizations' efforts to establish a counter-institutional position within an existing organizational field of real estate development as a foundation for producing a functional heterarchic (multi-lateral) governance structure. Their experiences demonstrate that an examination of efforts to realize just outcomes in urban development processes cannot be separated from the analysis of institutional structure.

COMMUNITY ORGANIZATION AND THE DISTRIBUTION OF POWER IN PLANNING

With regard to the distribution of power in urban development processes, there is a tension within the urban planning field between efforts to achieve more equal distribution of economic resources and more equal distribution of decision-making power.[2] The communicative rationalist stream of planning theory generally foregrounds equal participation in decision-making as a prerequisite for just economic redistribution (e.g., Forester 1989; Healey 1992a) while political-economic theorists generally emphasize the need for a reorganization of economic structure before democratic participation can be truly effective (e.g., Harvey 1996). Susan Fainstein's Just City formulation (see her chapter in this volume), which articulates philosophical and practical justifications for the prioritization of justice as a measure of urban development, focuses on Martha Nussbaum's (2000) "capabilities approach" as a way of theoretically mediating between these two emphases. While all of these perspectives have given some attention to the questions of institutional form (e.g., Healey 1999; MacLeod and Goodwin 1999), a specific focus on the organizational level[3] has been generally left out and this is especially the case within the more philosophical discussions around the Just City.

Distribution of economic resources and decision-making power at the organizational level has been a particularly salient issue for local community groups since the rise of the community development field in the 1960s. Many contemporary community development organizations in the United States can trace their roots to radical calls for local control and self-determination articulated most clearly by the Black Power movement and others in the 1960s (see Ture and Hamilton 1967; Foner 2002; DeFilippis 2004). Since then, however, neoliberal government initiatives emphasizing entrepreneurialism, deregulation and smaller government effectively appropriated the goal of local control and transformed it into devolution of responsibilities from the federal to state and local governments and to civil society organizations, including community developers. Though claiming to increase local democracy and individual liberty, neoliberal devolution ultimately decreased local control as it served primarily to reduce government's redistributive role at every scale (see Harvey 2005). As neoliberal political-economic theory took hold, the mainstream community development field and its primary organizational form, the Community Development Corporation (CDC), became increasingly entrepreneurial and professionalized as well as separated from its initial base of grassroots organizing and confrontational advocacy (Vidal 1992; Stoecker 1997; Vidal and Keating 2004).

This historical trajectory has meant that the challenge for community-based groups seeking more just urban environments in the context of shrinking government responsibility is not only to realize locally-based participatory democratic forms of urban governance, but also to find ways of leveraging the remaining redistributive power of the state in the name of the insurgent agendas that arise from such processes. At least in part, these insurgent agendas are formulated by organizations working within disinvested and declining communities during periods when there is little interest in the area from capital investors (see, for instance, the experiences of the Dudley Street Neighborhood Initiative detailed in Medoff and Sklar (1994)). Under such conditions of resource abandonment, the local organizations that remain are relatively free to determine the direction of development. As these groups succeed in improving their neighborhoods and market conditions change, however, developers and growth coalitions gradually reinvest in the area and threaten the self-determined status of local groups (Downs 1981: 75; Bluestone and Harrison 1982: 87). John Mollenkopf (1981: 331) has summarized this conflicted interdependence between capital and community. He writes, "Most evident is a cycle of growth and conflict in which the accumulation process leads to a growth in communities which it [capital] ultimately finds to be an impediment to further expansion."

As the case examined here demonstrates, community-development agendas formed during the periods of disinvestment that allow relative self-determination often require new strategies to effectively challenge private market interests during periods of reinvestment. As one activist interviewed in the course of our research stated, these strategies are "about changing the way that the city makes decisions about the land and the projects in our communities and . . . giv[ing] us a voice at the table." For this activist, real change is about leveraging community organization against existing state power in order to affect private development. While this leveraging has long been a strategy of community groups, academic research, real estate development networks and community tactics have all evolved since the early rise of CDCs referenced above. This evolution requires that the interorganizational mechanisms for directing reinvestment be continually reexamined.

Neoliberal devolution and changing investment strategies combined with the effects of urban environmental degradation require an increasingly complex organizational field[4] in order to manage brownfield redevelopment processes. Such complexity creates opportunities for innovation in urban development policy (Doak and Karadimitriou 2007: 210) as the case will show. However, in order to make that innovation serve insurgent community interests, mechanisms for altering existing network dynamics, and especially for directing the actions of highly connected entrenched organizations, are required. Urban development networks exemplify Perrow's (1991: 726) observation that "stratification within organizations and between them becomes the central determinant of our class system." How and if the intertwined race and class inequalities apparent within such stratified networks of organizational relations can be challenged by insurgent interests using strategies of institutional change is a central concern for those seeking to use brownfield redevelopment to further the community development agenda and for those seeking to create a Just City.

THE CASE: THE BROWNFIELD OPPORTUNITY AREA (BOA) LEGISLATION AND THE BRONX RIVER

The concentration of contaminated, abandoned land in many low-income neighborhoods such as the South Bronx in New York City is one of the most concrete representations of urban injustice. The areas that bore the brunt of the environmental degradation accompanying nineteenth- and twentieth-century urbanization and the development of industrial capitalism are now littered with vacant lots that are unremediated environmental hazards. It seems clear that in a Just City this land would at least be cleaned up and protected from re-contamination. Less clear though, is the question of who should have the power to decide the substance and shape of any development that takes place on the restored land.

This is especially the case in New York, where years of legislative impasse made it one of the last states in the country to pass a comprehensive law regulating the cleanup and redevelopment of contaminated former industrial sites. In 1994, when the legislative gridlock looked likely to extend indefinitely, the New York State Department of Environmental Conservation (DEC) created a "Voluntary Cleanup Program" under which property owners agreed to clean up their land in return for assurances that they would not be sued by the department for future remediation (Hu 2003: B1). The law, however, did not spell out safe pollution levels. The DEC promulgated regulations saying that the goal should be to return the land to a "pristine state," but that goal and more specific guidelines for each chemical were rarely met. Instead, state officials negotiated deals with each polluter on a case by case basis (McKinley 2002: B1). This let major polluters off the hook and forced community organizations to engage in long, costly battles to ensure the safe cleanup of toxic sites.

A case in point is Starlight Park, a public park along the Bronx River. In 2000, Starlight Park was found to be heavily contaminated from the remains of a coal gasification plant that had been operated on the site by a predecessor of the Con Edison utility company from the 1880s until the 1920s. A local community organization, Youth Ministries for Peace and Justice (YMPJ), spent the following six years mobilizing residents to make sure that the DEC and the New York City Parks Department (the current landowner) held Con Edison to commonly accepted minimal standards for remediation. The cleanup finally began in the fall of 2006, and included reparations paid by Con Edison to the Parks Department for construction of the park. The rare victory that the community group was able to win in this case came only through constant struggle, and such victories have been the exception, not the rule, for the cleanup program. The voluntary cleanup program essentially let polluters avoid responsibility at minimal costs, resulted in incomplete remediation and had no structures which ensured residents that future development would not repeat the same destructive cycles. When comprehensive cleanups were conducted it was only because of the vigilance and advocacy of local organizations that were able to bring public attention and political power to bear on the landowners and polluters.

Frustrated by fighting continuous battles waged on a site-by-site basis similar to that conducted over Starlight Park, environmental justice groups statewide began to work together to devise a more effective approach. In 1998, funded by the Rockefeller Brothers Foundation, New York environmental justice groups came together with representatives of some of the large polluting industries, city and state officials and other stakeholders to form a Brownfields Coalition to draft and advocate for more comprehensive, rational statewide legislation. As part of the effort, the Environmental Justice groups advocated for a bill that would provide public funding and support for

Figure 10.1 Ajamu Kitwana, YMPJ staff, and Anthony Thomas, YMPJ youth member, inspecting progress in the remediation of Starlight Park

community groups or municipalities to perform area-wide surveys of brownfields and engage in participatory community processes to establish residents' priorities for local redevelopment. In 2003 after lobbying and advocacy by local and state-wide organizations, the state senate and assembly passed the Brownfield Opportunity Areas (BOA) legislation based on the model the groups had drafted along with the Brownfield Cleanup Program (BCP) bill which provided tax-credits to developers for brownfield redevelopment.[5,6]

One of the early groups to receive a BOA grant was a coalition centered around three Bronx organizations founded by local residents and led by women of color from the neighborhood. The groups were Youth Ministries for Peace and Justice (YMPJ), a membership-based community group organizing young people for environmental justice in the Bronx River and Soundview neighborhoods; Sustainable South Bronx (SSB), a non-profit environmental justice solutions corporation developing economically sustainable projects informed by community needs; and The Point, a non-profit community development corporation focused on youth development and the cultural and economic revitalization of the Hunts Point neighborhood, with a strong focus on the arts. These three community-based groups worked in coalition with two technical assistance providers: The Pratt Center and the Bronx Overall Economic Development Corporation (BOEDC).

The Southern Bronx River Waterfront BOA (SBRW BOA), as the coalition is known, is currently concluding its survey of brownfield sites and conducting workshops with local community members. Though any conclusions drawn are necessarily incomplete, their experiences point to some of the possibilities and limitations of the approach these actors have taken to reforming the institutional model of urban development founded upon notions of justice established by the Environmental Justice movement. For the analysis that follows, we supplement two years of participant observation with YMPJ completed by one of the authors with twelve semi-structured interviews performed over a two-month period in 2007. We interviewed organization leaders in all of the community and technical assistance groups involved with the SBRW BOA as well as city officials identified by community group leaders or by job description as connected to the actions of the SBRW BOA. Each interview lasted between one and two hours and focused on the type and nature of interactions that each group has with city, state, private development and community organizations as a result of forming the BOA. As well, we discussed the goals of the group, the greatest challenges faced by the BOA, alternative institutional structures for brownfield development and additional issues raised by individual interviewees were discussed. Common themes raised within these interviews are contextualized through the knowledge gained from participant observation in the analysis that follows.

Figure 10.2 A planning meeting for the SBRW BOA

ENVIRONMENTAL JUSTICE, SELF-DETERMINATION AND REDISTRIBUTION

The actions of the SBRW BOA are deeply rooted in the model of social justice developed by the Environmental Justice (EJ) movement in the United States. In contrast to the larger environmentalist movement, which is often seen as dominated by liberal conservation groups, environmental justice activists begin from a recognition that the immediate human consequences of pollution and environmental destruction in the US are felt most directly by poor people of color who live in neighborhoods where environmental hazards and contaminants are disproportionately concentrated (see US GAO 1983; Commission for Racial Justice 1987). For example, attempts to site a landfill for PCB contaminated soil in largely African-American Warren County, North Carolina in 1982 sparked large, highly publicized protests and gave a name to the concept of environmental racism—the targeting of communities of color for waste disposal and polluting industrial activity (Bullard 1990). Shortly thereafter, dispersed EJ struggles coalesced through the groundbreaking 1991 National People of Color Environmental Leadership Summit and revived earlier emphases on self-determination and grassroots organizing to realize redistributive urban development outcomes (for more on the development of a national Environmental Justice Movement, see *inter alia* Bullard 1993, Adamson 2002, Bullard *et al.* 2004).

The EJ movement's founding principles call for "the fundamental right to political, economic, cultural and environmental self-determination of all peoples" and "demands the right to participate as equal partners at every level of decision-making, including needs assessment, planning, implementation, enforcement and evaluation."[7] The principles of Environmental Justice are similar to the Just City formulation in that both return to the fundamental philosophical and moral underpinnings of shared understandings of justice in order to mobilize support for spatially conscious social change. The two approaches differ in that, while the Just City draws on Western philosophy, the EJ movement emphasizes the sacredness of the earth and the roots of collective commitments to justice in our diverse spiritualities. The EJ principles resist commodification, either through the payment of community benefits before development or monetary damages after the fact, seeking instead the transformation of our relations to one another and to the earth. The transformations of these relations in the EJ model must begin at the grassroots, from the particular context of specific everyday lives. Justice for EJ organizers then cannot be defined abstractly, but must be achieved through self-determination of marginalized communities at the local level.

The experiences of residents of marginalized communities of color in the U.S. have repeatedly highlighted the ways in which different oppressions are intertwined. Accordingly, the EJ movement combines emphases on environmental and social justice, understanding the environment to encompass the totality of life conditions, including air and water and access to open spaces and recreation, as well as working conditions and wages and the quality of housing, education, health care and transportation. In order to engender greater equality and health in cities and simultaneously maintain local control, EJ organizations have developed a structure of small community-based groups that mobilize active memberships at the local level and form networked coalitions which sometimes then work in dialogue with city and state agencies. This locally networked approach can be seen as an attempt to turn the neoliberal promotion of devolution against itself and leverage the powers of local governments to achieve more just redistribution of resources. The BOA, in the words of one of its staff members, "is about creating a long term change in the nature of the relationships between decision-makers [in urban development] and community members." Success for the Bronx EJ organizations studied here would mean simultaneously enabling greater participatory democracy via an accessible infrastructure of local organizations and improved environmental health and economic redistribution gained via inter-connections with public-sector agencies and the regulations over private market actors that they direct. Such a strategy acknowledges that in practice a Just City requires that both the redistribution of economic resources and decision-making power be resolved together.

The South Bronx, with its recent redevelopment (e.g., large-scale projects such as the Bronx River Greenway, Bronx Terminal Market, and Yankee Stadium) and incipient gentrification (e.g., its rebranding as "SoBro"), has been a key area for environmental justice organizing. Despite a recent spike in real estate development, the area continues to have one of the highest unemployment rates in the nation and includes some of the U.S.'s poorest congressional districts. It also contains hundreds of contaminated brownfield sites. The correlation of race, class, and high levels of contamination is not accidental. Cycles of industrialization and deindustrialization and investment and disinvestment of capital have followed the logic of profit maximization and political expediency, which suggests that pollution should be concentrated in those areas with the lowest property values and wages and the least political resistance (see e.g., Squires 1994; Harvey 1997).

Thus brownfields are revealed not so much as a liberal issue of "sustainability," but as a product of social processes that benefit some groups while adversely affecting others. After more than a decade of struggle, Bronx EJ groups won tremendous victories in the allocation of land and funding for the creation of the Bronx River Greenway, a network of public parks providing access to the water and recreational and economic development opportunities for the neighborhoods along the Bronx River that currently have a deficit of open space but an abundance of serious environmental health hazards. However, just as groups were celebrating the groundbreaking of the greenway, a city-commissioned consultancy report was leaked identifying the Bronx River as a site for rezoning of existing industries and brownfields for high-end residential use (see Garvin Report 2006). As land prices have increased, South Bronx brownfields are becoming a locus of conflict between, on the one hand, the interests of long-standing community-based organizations seeking to maintain what hard won self-determination they have with regard to neighborhood growth and, on the other hand, the interests of private developers and property owners seeking to maximize the profits that inner city post-industrial land offers.

In this context, a director of one of the BOA groups summarized her understanding of environmental justice as the recognition that if you do not want to repeat the disinvestment and environmental destruction that tore apart the South Bronx:

> You have to address the inequalities that created the problem in the first place and yes that is disparate environmental burdens, but even more so it is communities that didn't have the knowledge and the power to make sure that this environmental contamination doesn't happen to begin with. For some people coming in and giving me a park or a clean river or removing a highway is enough, it's eliminating a burden or giving me a benefit. But the spirit of environmental justice, which I learned from people who came before me, has always been about this power and knowledge and self-determination . . . For me, the BOA is extraordinary in as much as it is . . . about . . . ensuring that people have a say in their community.

Redefining community as counter institution

The SBRW BOA has sought to create the systemic power that community members need to ensure that contamination does not happen again and that redevelopment helps to ameliorate inequality by first altering the institutional definition of community. BOA groups seek to infuse themselves into the land-use decision making process, by expanding the formally recognized organizational identity for local neighborhoods. As one BOA member asks: "How do we define who the community is? Who gets to decide that? That's a major issue with the BOA."

Existing outlets for community representation in New York City grew from advocacy for greater participation in land-use decisions in the 1950s and 60s and led to the creation of Community Boards. The Boards represent neighborhoods in the Uniform Land Use Review Process (ULURP) for zoning changes, special permits modifying zoning controls, site selection for capital projects, urban renewal plans, and the disposition or acquisition of city owned property. Community Board members are all appointed by the borough presidents and largely serve at their whim. As a result, rejection of proposals is rare and the Boards have taken on a predominantly system-affirming role in land-use processes.[8]

The community-based organizations that comprise the SBRW BOA often work with their Community Boards, but they differentiate themselves from these appointed bodies. The Bronx River/Soundview Community Board, whose district covers much of the area the SBRW BOA is active in, is largely made up of home and business owners with higher incomes and educational levels than the neighborhood median, and with some connection to the Borough President or local City Councilors. Members of the grassroots BOA groups are more representative of the majority of neighborhood residents who are lower-income renters, the many residents who are more recent immigrants, as well as young people excluded from the formal electoral process.

This close connection to the segments of the population with less formal representation in land-use processes often generates a willingness to challenge those in existing positions of power with regard to development and in turn has provided grassroots organizations with a base of support that is distinct from the Community Boards—their legitimacy has historically been based upon their oppositional stances, not their connections to decision-making. In conjunction with their members and other community residents, the BOA groups have consistently challenged proposals that represent the neighborhood as a clean slate in need of market-rate housing and big box retail and emphasized instead the need for local economic development that supports existing small businesses and creates healthy jobs with decent wages for local residents, as well as housing that guarantees permanent affordability. For example, instead of advocating the wholesale eviction of auto-repair shops and other small industries along the Bronx River as the Garvin Report did, the BOA groups have been researching ways these small businesses can operate more sustainably for their employees and the neighborhood, perhaps by accessing state and federal financing to become leaders in servicing alternative fuel-based transportation.

The counter-positions that emerge from the connection of these groups to the majority of neighborhood residents excluded from formal decision-making processes and which define the identity of the EJ organizations in the Southern Bronx Waterfront BOA are consistent with the principles of environmental justice in that they reflect the idea that the just exercise of power is characterized by the participation of those most affected and most marginalized in decision-making. The approach explicitly recognizes the interdependence of life as well as the right to self-determination and is congruent with Iris Marion Young's (2000) theorizations about the conditions that can create inclusive democratic communication in spite of structural inequality and cultural difference. Young has called this approach a "democratic theory for unjust conditions" that establishes the "institutional conditions for promoting self-development and self determination of a society's members" (Young 2000: 33).[9] As one interviewee said:

> The hope for the BOA was to have not just one group pursuing its own agenda or projects but a strong coalition of groups that come from different perspectives and ideologies but are united in their key concerns about the harm that speculators are doing by tying up key sites and

> the need for more city action on brownfields. The hope was that these groups could reach out to other stakeholders—to other regular folks in the neighborhood, to local businesses, churches and tenants' associations—and with a unified vision backed with state money become recognized by the city and the state agencies making land-use decisions.

Such just exercise of power within the context of urban development requires that those most affected by development have a mechanism at their disposal for empowering their own plans for their communities as well as the power to make those plans actually guide development. The BOA seeks to create this mechanism for planning and empowerment via an adaptation of the existing organizational field of land-use decision-making for brownfields. BOA groups, along with city and intermediary organizations, are lobbying the New York state legislature to amend the brownfield cleanup law such that receipt of a percentage of the state brownfield tax credits available to developers is contingent upon approval of the proposed project by the local BOA based upon their area-based strategy for redevelopment.[10] This "linkage" requirement would be a financial carrot encouraging landowners and developers to work with BOA groups and opening some of the state's economic capital to contestation and negotiation within the organizational field. In the words of one interviewee: "We hope that the linkage would help create a meaningful connection with land owners and developers and encourage them to collaborate with BOA groups, encouraging them to buy into the process for their own goals and being able to access money from the state as a part of it."

In essence, the linkage requirement is an effort to give institutional power to the alternatively defined community that the BOA represents. This institutional power is not only provided by economic capital from state-level government agencies. BOA groups are looking to the city to back up the carrot of extra money from the state with the stick of withholding city development approvals unless private brownfield developers first consult with BOAs for the project they are proposing.[11] "Almost every major development site within the BOA area would need a zoning change to maximize its potential," one of the BOA groups' staff members pointed out,

> and even the most hardened neoliberal would acknowledge that government needs to be a player in brownfield redevelopment because of the contamination and infrastructure issues and the risks and costs involved . . . That should be a point of entry for people in all parts of the city that have brownfields to say, 'if public money is going to be spent, a commensurate public benefit must be delivered.' And the planning process is about what are those benefits.

This somewhat incongruous movement by the BOA groups *toward* institutional power as a means of challenging the existing institution embodies what has sometimes been referred to as a "counter-institutional" position (see Fainstein this volume; Marcuse this volume). The counter-institution has been described in reference to Jacques Derrida's work as a "with–against" movement—at once a part of the established order while simultaneously a seed of change within this order (Wortham 2006: 1–24). Beyond simply a reform effort, counter-institutions supplement the existing institutional context "by more adequately fulfilling its goals, that is, the goal of collective working-together on the basis of some kind of consensus. At the same time the counter-institution brings into the open what keeps the institution from ever fulfilling its goals" (Miller 2007: 284). Using tools such as "counterpublic discourse" (see Wolf-Powers this volume), a counter-institution reveals institutional shortcomings by maintaining an alternate vision for existing institutional forms towards which it is working, a vision which "must respond . . . to an infinite demand for justice" (Miller 2007: 292).

In this sense, the counter-institution that the BOA creates is utopian realist—utopian in that it holds up an alternate vision for organizations that are relevant to land-use decisions and realist in that the alternate vision builds upon the existing organizational structure. This approach has a great deal of resonance with the means by which Henri Lefebvre sought to realize his "right to the city" concept. Lefebvre argues that planners (and everyone else) should always have in mind a utopian vision of what urban space can be, but root their actions in what urban space is (Lefebvre 1996). Thus, the BOA groups' work to realize their vision of justice by re-ordering organizational relations and capturing some of the state's redistributive power in order to give agency to those seeking a redress of fundamental inequality demonstrates the crucial role of institutional development and interorganizational mechanisms as means for realizing the "right to the city" and creating a more just urban environment.

EMPOWERING HETERARCHIC GOVERNANCE: LESSONS FROM THE LORAL SITE

The goal of creating a counter-institution premised on inclusive decision-making in urban development processes requires that community-based organizations not only advance a definition of community within the land-use decision-making process different from the appointed community boards in order to legitimate their insurgent agenda, but also that they effectively connect this agenda to larger fields of policymaking. In order to do so, they must incorporate their counter-institutional position into a viable heterarchic governance model. Heterarchic governance has been described most prominently as "neither market nor hierarchy" (Powell 1990). It is a model applied to efforts which rely heavily on dialogue between organizations and individuals to achieve multi-lateral decision-making processes that downplay both hierarchic power and individualistic action. The concept has been applied to the study of regional economies, industrial districts, transitioning national régimes, and urban planning processes, among others (see Powell 1990; Stark 1996; Jessop 1998; Stone 2006).

Importantly for the case examined here, heterarchic governance is especially suited for situations where established institutional arrangements are in the process of being altered (Stark 1996; Jessop 1997). In processes of urban development, it characterizes periods where private market actors give up some of their autonomy in decision-making and state actors give up some of their top-down authority. In order to bring this about, Jessop (1998: 36) argues, "the 'added value' that comes from partners combining resources rather than working alone" *must* be evident to all involved. From this, an inter-organizational capacity that is greater than that of any individual member arises. However, Jessop emphasizes the dangers of uncritically celebrating heterarchic forms of governance since they do not change market principles (1998: 39). Rather, it is simply a new, if more complex and wider, arena where the antagonisms created by competition for capital are expressed. The earliest test of the SBRW BOA at a large brownfield known as the "Loral Site" demonstrates this point well.

The Loral Site is an abandoned industrial waterfront manufacturing facility near the intersection of the Bronx River and the East River, where the Loral Corporation had formerly manufactured electronic components for the Air Force during the Cold War. The site is immediately adjacent to several low- and middle-income Mitchell-Lama subsidized housing developments mostly occupied by long-term residents. The majority of the site is privately owned and the revelation in 2006 that it was for sale sparked a contested debate about its future described by several of our interviewees, who provided the details below.

Tenants in the housing development got early word of the sale and contacted Sustainable South Bronx (SSB) and Youth Ministries for Peace and Justice (YMPJ). SSB discovered that Peter Fine, a New York based developer, was the leading contender for purchasing and developing the site. Upon researching Fine's track record, the Pratt Center found evidence of poor labor practices, low levels of affordability and poor quality construction. Though the BOA organizations had not yet begun the Brownfield Opportunity Area community planning process and no legislation creating the "linkage" described above had been passed, the groups drafted a letter from the BOA and sent it to the site's owner, potential purchaser and local elected officials. They asserted the planning function of the BOA, identified the Loral Site as a key part of their area and stressed that new development must respond to community concerns. Shortly thereafter, the deal with Peter Fine fell through. The exact reason why Fine backed out was not known by our interviewees, but it seems that the SBRW BOA played at least some role.

The next potential purchaser of the Loral Site came with the general approval of the community groups. Carlton Brown of Full Spectrum Development is an African-American developer with strong environmental health and environmental justice credentials who was already known to leaders in several of the BOA organizations. Though he had not worked with all of the BOA groups and some remained skeptical, Brown gave early recognition to the BOA approach and substantially engaged BOA groups in the project development phase. Full Spectrum's staff arranged several face to face meetings with all the stakeholders and delivered prompt answers to questions posed by the group members about levels and permanency of affordability, public access to open space and the river, mitigation of runoff into the river and other green design elements. Soon, though, the deal with Brown fell through when Apollo Capital, the financier of the nearby state subsidized Mitchell-Lama houses encouraged the tenants' associations not to give up the covenant they hold over the height of future development on the Loral Site.

The tenants' associations largely opposed more housing construction and workshops and discussions with the BOA groups failed to convince them that Brown's proposal was the best option. This setback effectively undid Brown's ability to develop his vision of affordable housing and public open space along the river. As it turns out, Apollo Capital may have had motives of their own in encouraging the holdout since they were rumored to be the finance group behind a potential new purchaser.

This example demonstrates both the alternate institutional vision for brownfield development that the SBRW BOA seeks as well as the indivisible nature of just urban development outcomes and a functional heterarchic governance model which can incorporate counter-institutional positions. The BOA groups must connect with the larger field of policymaking within the context of an existing entrenched interconnection between the state's political capital and private developers' economic capital.[12] Such a pre-condition of the organizational network has a strong effect upon the ability of these groups to realize just outcomes. Their failure to affect the outcome of the Loral Site is premised upon the disadvantaged organizational position of the BOA groups and the lack of a functional heterarchic governance structure, which would enable the recognition of added value that comes with BOA consultation. This recognition is enabled in part from the mechanical creation of value demonstrated in the linkage and approval requirements sought, but also from the residual trust and recognition of roles that comes about with the incorporation of these mechanisms into a re-formed network of brownfield development actors.

While a full network analysis cannot be developed here, an outline of a few characteristics of the organizational field that the SBRW BOA is working in demonstrates some of the as yet unmet demands of heterarchic governance highlighted by actions around the Loral Site. The field of organizations concerned with brownfield redevelopment in the South Bronx is made of community-based groups, political intermediaries, city and state agencies, and private development interests. Connections between these groups result when organizations meet formally to plan and develop projects, share and hire staff from among one another's ranks, or maintain informal open lines of communication to sort out and prioritize goals.[13]

Our analysis of the network shows that three of the five organizational types, specifically city, state, and intermediary organizations, are connected in some way to all other organizational types in the network and city agencies have the strongest connections, and thus the central position, with all other groups. The only two organizational types not connected to all others are community groups and private development interests. Our interviews indicate that these two organizational types are not yet connected to one another in any substantial way with regard to brownfield development processes in the South Bronx. The disregard with which Apollo Capital and the new purchaser of the Loral Site treated the BOA groups demonstrates this structural hole, or missing link, between community groups and private developers. While Full Spectrum and the BOA groups attempted to span these two sides in the major contestation within the organizational field that the SBRW BOA is trying to affect—that is the struggle between non-state institutions for access to public capital, largely political and economic, needed to get projects built in the South Bronx—they were ultimately thwarted because the mechanisms for establishing trust[14] had not been developed sufficiently to reach the tenants' organizations who remained suspicious. As well, the city government played no active part in bringing them together, such as mediating between the stakeholders or publicly recognizing the BOA in ways that would strengthen its legitimacy and negotiating power.

Intermediary organizations are working to create such mechanisms by aligning themselves closely with community groups, even while reaching out to state, city, and private market organizations to form bridges across structural holes in the organizational field. In the words of one interviewee, the primary intermediary for SBRW BOA, New Partners for Community Revitalization (NPCR), "speaks everyone's language." The interviewee was referring to the fact that NPCR staff have technical engineering, planning and legal backgrounds, spend time advocating for new policies at the state capital and have a long-established relationship with local community groups allowing them to communicate freely with all organizational types in the network. They are "political buffers" who can go out on a limb on an issue without any specific community group having to take the risk of being attached to it.[15] NPCR navigates this politically sensitive position by mirroring the structural role that the city plays within the organizational field but, where city agencies' connectedness within the network has been stronger with private development interests, NPCR focuses its connectedness on strengthening the position of community organizations (see Figure 10.3).

Ideally, this effort in conjunction with the ongoing outreach on the part of the BOA groups themselves will lead to a more balanced organizational position for BOA groups. This position would give them greater leverage in directing actions during circumstances such as the sale of the Loral Site because they would be connected to the active organizations and have access to an established base of resources that is understood by all actors within the network. Such balance is achieved by leveraging the connections created across structural holes against the fulcrum of state power as shown schematically in Figure 10.3. This outcome is premised on the "added value" of community

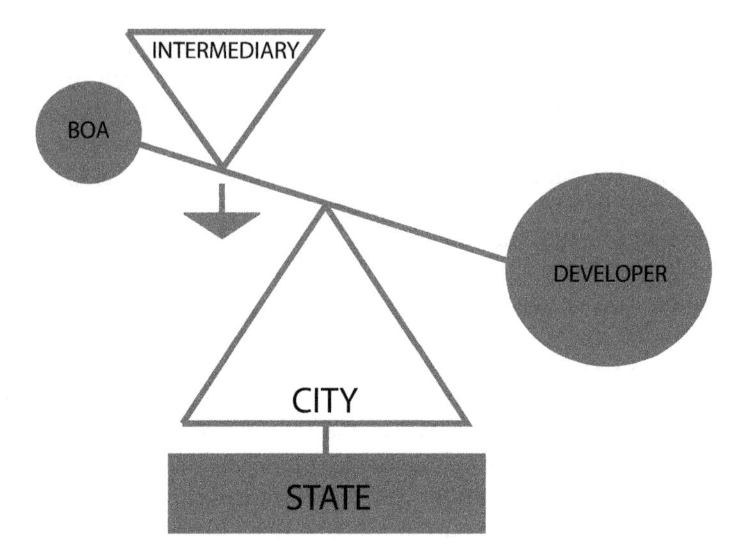

Figure 10.3 Relationship between the five organizational types, showing efforts to leverage state power for a more balanced position for community groups

organization involvement in the land-use decision-making being visible to all actors in the network. This is a key to establishing the alternate institutional form that the BOA counter institutions seek to bring about and it is the required condition for a functional heterarchic governance model.

CONCLUSION

As the Loral example demonstrates, despite the fact that land owners literally ran away from the idea of individual responsibility for the South Bronx during the long years of disinvestment in the 1970s and 1980s, the notion of individual development rights in search of the highest profits is now being asserted by those same developers as a universal value underpinning justice. Now that the groundbreaking Bronx River Greenway is well under construction and there is increasing market pressure on local renters and local landowners, how can cash-poor community groups maintain a level of self-determination for their constituents? Without the development of an empowered counter-position nurtured by a functional heterarchic governance structure, a strong emphasis on individual owners' property rights will be the sole value that guides development in the South Bronx. That is, the existing norms reinforced by the current structure of real estate development networks will remain.

Representing different segments of the community than those currently seen on local community boards, the BOA groups are articulating an insurgent planning agenda and seeking institutional power in the process. As planners, the BOA groups are not acting in the Habermasian sense of mediators seeking truth through ideal discourse, but as Gramscian organizers locating themselves within the institutional field and intervening strategically to create new forms of local accountability and more redistributive outcomes. These strategic interventions require negotiating the conflicts that emerge between the diverging definitions of values that different actors bring to the table and questioning the assumption that a universal value must be static. The "right to the city" that EJ groups seek is continually determined based upon local conditions and what the city is, but it is asserted within the context of a vision for the Just City which tells us what the city may be.

As Iris Marion Young (1990) proposes, socially just urban planning requires mechanisms for incorporating difference into decision-making processes, but as Susan Fainstein reminds us, it also requires a vision for what just urban development might be. Collectives of individuals such as the EJ groups can only negotiate their differences with other urban actors, however, when structural holes in organizational networks which define the major stratifications of power, often along race and class lines, have been bridged (unlike the case of the Loral Site). Without such connection, their dialogue has no outlet and their interests will likely be forced to become short term and insular, as the tenants' associations near the Loral Site demonstrated. The realization of a just city requires a vision for just outcomes that can inspire and motivate these bridges to be formed. A balanced organizational field premised on redistribution of organizational powers can serve as such a vision because it allows for counter-institutional perspectives, such as that of environmental justice, to be incorporated. Thus, such an institutional form addresses both process and product, and can be an essential aspect in articulating a model for the Just City.

NOTES

1 This chapter was equally co-authored. The order of names is alternated in successive published pieces by these authors.

2 See, for instance, Fraser (1995) on the related issue of redistribution and recognition, and Beauregard (1998) and Sandercock (2003) for an analysis of this tension in urban planning theory and practice.

3 For an especially astute analysis of the importance of the organizational level of analysis to the question of urban social justice, see Marwell (2007: 1–32). Marwell (2007: 7) argues that: "When community-based organizations work to improve the conditions of poor neighborhoods they essentially attempt to reshape economic and political fields."

4 The notion of organizational field is employed here in the sense that Marwell (2007: 3) defines it: "a set of organizations linked together as competitors and collaborators within a social space devoted to particular types of action." This notion is differentiated from an organizational network in that a network is comprised solely of collaborators. For more on the institutionalist perspective on organizational fields, see Dimaggio and Powell (1991).

5 For more about the conflicts and compromises involved in the legislation see Baker (2003); Bridger (2003); Siska (2004); Steinhauer (2005).

6 The BOA program mandates the Department of State and the Department of Environmental Conservation to provide financial and technical assistance to local governments or community organizations to conduct comprehensive assessments of brownfield sites. The BOA program sets out a three-phase process for community groups or municipalities that want to undertake brownfield studies: pre-nomination analysis, nomination study and implementation strategy. Groups or municipalities can then partner with the state to market the brownfield sites to developers for reuse consistent with the plan prepared by the community group. For more information see: http://www.nyswaterfronts.com/grantopps_BOA.asp

7 See the full "Principles of Environmental Justice" at: http://www.ejnet.org/ej/principles.html accessed 6/05/08. Other relevant principles adopted at the Summit state that environmental justice:

> 1) affirms the sacredness of Mother Earth, ecological unity and the interdependence of all species, and the right to be free from ecological destruction. 2) demands that public policy be based on mutual respect and justice for all peoples, free from any form of discrimination or bias. 3) mandates the right to ethical, balanced and responsible uses of land and renewable resources in the interest of a sustainable planet for humans and other living things . . . 5) affirms the fundamental right to political, economic, cultural and environmental self-determination of all peoples . . . 7) demands the right to participate as equal partners at every level of decision-making, including needs assessment, planning, implementation, enforcement and evaluation . . .

8 While the Community Boards were created to foster local participation and control over land-use, they have been increasingly critiqued for their lack of independence, as evidenced for instance by the politically motivated purging of members from Brooklyn Community Board Six who opposed the recent Atlantic Yards development plans and from Bronx Community Board Four who opposed the taking of public parks for a new Yankee Stadium. While some of New York's Community Boards, such as East Harlem's, are known as participatory bodies that do actually ensure that residents have a meaningful say in local development, the level of democracy and effectiveness in others is less certain.

9 Young premises her argument on the claim that deliberative democracy, along with inclusion and political equality "increases the likelihood that democratic decision-making processes will promote justice" (Young 2000: 6).

10 Between the completion of this chapter and its publication, the New York State legislature passed bill S.8717/A.11768, which provides a 2 percent Brownfield Tax Credit bonus for projects built in and conforming with BOA plans. This makes future analysis of the impacts of this "linkage" program especially pertinent with respect to the interorganizational dynamics that effect use of the linkage bonus.

11 The Mayor's Office in New York City announced on June 9, 2008 that it would create a new Office of Environmental Remediation. One of the stated goals of the office is to "enhance the brownfield opportunity area program to provide community groups the planning resources that they need." This could be an opening for community groups to get the backing that they need from the city in terms of project approvals.

12 The connection between local bureaucracy and private development interests has long been catalogued within urban political science. See, for example, Hunter (1953); Logan and Molotch (1988); Stone (1989); Dahl (1961).

13 Culled from interview data.

14 For more on the role of trust in networks of organizational governance, see Powell (1996); see also Granovetter (1985); Ring (1997).
15 From interview notes.

REFERENCES

Adamson, J. (2002) *The Environmental Justice Reader*, Phoenix: University of Arizona Press.
Baker, A. (2003) "Senate approves plan to clean polluted sites, ending 10-year impasse," *New York Times*, September 17: B6.
Beauregard, R. (1998) "Writing the planner," *Journal of Planning, Education and Research*, 18: 93–101.
Bluestone, B. and Harrison, B. (1982) *The Deindustrialization of America: Plant Closings. Community Abandonment, and the Dismantling of Basic Industry*, New York: Basic Books.
Bridger, C. (2003) "ECIDA opposes brownfield measure," *Buffalo News*, September 11: C9.
Bullard, R. (1990) *Dumping in Dixie: Race, Class, and Environmental Quality*, Boulder, CO: Westview.
—— (1993) *Confronting Environmental Racism: Voices from the Grassroots*, Boston: South End Press.
Bullard, R., Johnson, G., and Torres, A. (2004) *Highway Robbery: Transportation Racism and New Routes to Equity*, Boston: South End Press.
Commission for Racial Justice (1987) *Toxic Wastes and Race in the United States*, New York: United Church of Christ.
Dahl, R.A. (1961) *Who Governs? Democracy and Power in an American City*, New Haven, CT: Yale University Press.
DeFilippis, J. (2004) *Unmaking Goliath: Community Control in the Face of Global Capital*, London: Routledge.
DiMaggio, P.J. and Powell, W. (1991) *The New Institutionalism in Organizational Analysis*, Chicago and London: University of Chicago Press.
Doak, J. and Karadamitriou, N. (2007) "(Re)development, complexity, and networks: a framework for research," *Urban Studies*, 44(2): 209–229.
Downs, A. (1981) *Neighborhoods and Urban Development*, Washington, D.C.: The Brookings Institution.
Foner, P. (ed.) (2002) *The Black Panthers Speak*, Cambridge, MA: Da Capo Press.
Forester, J. (1989) *Planning in the Face of Power*, Berkeley: University of California Press.
Fraser, N. (1995) "From redistribution to recognition? Dilemmas of justice in a 'post-socialist' age," *New Left Review*, 212: 68–93.
Garvin Report (2006) *Visions for New York City: Housing and the Public Realm*, prepared by Alex Garvin and Associates for the New York City Economic Development Corporation, available at: http://www.streetsblog.org/wp–content/uploads/2006/08/Garvin_Report_Full.pdf (accessed on June 25, 2008).
Granovetter, M. (1985) "Economic action and social structure: the problem of embeddedness," *American Journal of Sociology*, 91(3): 481.
Guy, S. and Harris, R. (1997) "Property in a global risk society: towards marketing research in the office sector," *Urban Studies*, 34(1): 125–140.

Harvey, D. (1996) "On planning the ideology of planning," in Fainstein, S. and Campbell, S. (eds) *Readings in Planning Theory*, Oxford: Blackwell.
—— (1997) *Justice, Nature and the Geography of Difference*, Oxford: Blackwell.
—— (2005) *A Brief History of Neoliberalism*, Oxford: Oxford University Press.
Healey, P. (1992a) "Planning through debate: the communicative turn in planning theory," *Environment and Planning B: Planning and Design*, 23: 217–234.
—— (1992b) "An institutional model of the development process," *Journal of Property Research*, 9: 33–44.
—— (1994) "Urban policy and property development: the institutional relations of real estate in an old industrial region," *Environment and Planning A*, 26: 177–198.
—— (1999) "Institutional analysis, community planning, and shaping places," *Journal of Planning Education and Research*, 19: 111–121.
Hu, W. (2003) "A sure deal on brownfields? Don't forget this is Albany," *New York Times*, June 24: B1.
Hunter, F. (1953) *Community Power Structure: A study of decision makers*, Chapel Hill, NC: University of North Carolina Press.
Jessop, B. (1997) "The governance of complexity and the complexity of governance: preliminary remarks in some problems and limits of economic guidance," in Amin, A. and Hausber, J. (eds) *Beyond Market and Hierarchy: Interactive Governance and Social Hierarchy*, Lyme, UK: Edward Elgar.
—— (1998) "The rise of governance and the risk of failure: the case of economic development," *International Social Science Journal*, 50(155): 29–45.
Lefebvre, Henri (1996) "The right to the city," trans. in Koffman, E. and Lebas, E. (eds) *Writings on Cities*, Oxford: Blackwell Publishing.
Logan, J. and Molotch, H. (1988) *Urban Fortunes: The Political Economy of Place*, Berkeley: University of California Press.
MacLeod, G. and Goodwin, M. (1999) "Reconstructing an urban and regional political economy: on the state, politics, scale and explanation," *Political Geography*, 18(6): 697–730.
Marwell, N. (2007) *Bargaining for Brooklyn: Community Organizations in the Entrepreneurial City*, Chicago and London: University of Chicago Press.
McKinley, J. (2002) "Impasse in Albany stalls financing for superfund," *New York Times*, July 3: B1.
Medoff, P. and Sklar, H. (1994) *Streets of Hope: The Fall and Rise of an Urban Neighborhood*, Boston: South End Press.
Miller, J.H. (2007) "'Don't count me in': Derrida's refraining," *Textual Practice*, 21(2): 279–294.
Mollenkopf, J. (1981) "Community and Accumulation," in Dear, M. and Scott, A.J. (eds) *Urbanization and Urban Planning in Capitalist Society*, New York: Methuen.
Nussbaum, M. (2000) *Women and Human Development: The Capabilities Approach.* Cambridge: Cambridge University Press.
Perrow (1991) "A society of organizations," *Theory and Society*, 20(6): 725–762.
Powell, W. (1990) "Neither market nor hierarchy: network forms of organization," *Research in Organizational Behavior*, 12: 295–336.
—— (1996) "Trust-based forms of governance," in Kramer, R. and Tyler, T. (eds) *Trust in Organizations: Frontiers of Theory and Research*, Thousand Oaks, CA: Sage Publications.
Ring, P. (1997) "Process facilitating reliance on trust in inter-organizational networks," in Ebers, M. (ed.) *The Formation of Inter-Organizational Networks*, Oxford: Oxford University Press.

Sandercock, L. (2003) *Cosmopolis II: Mongrel Cities of the 21st Century*, London: Continuum.
Siska, D. (2004) "The brownfields breakthrough," *Foundation News*, May/June, 45.
Squires, G. (1994) *Capital and Communities in Black and White: the intersections of race, class, and uneven development*, Albany, NY: State University of New York Press.
Stark, D. (1996) "Recombinant Property in East European Capitalism," *American Journal of Sociology*, 101(4): 993–1027.
Steinhauer, J. (2005) "A cleanup that's easier legislated than done," *New York Times*, December 4: A47.
Stoecker, R. (1997) "The CDC model of urban development: a critique and an alternative," *Journal of Urban Affairs*, 1: 1–22.
Stone, C.N. (1989) *Regime Politics*, Lawrence, KA: University Press of Kansas.
—— (2006) "Power, reform and urban regime analysis," *City and Community*, 5: 1.
Ture, K. and Hamilton, C.V. (1967) *Black Power: The Politics of Liberation*, New York: Vintage Books.
U.S. Government Accounting Office (US GAO) (1983) *Siting of hazardous waste landfills and their correlations with racial and economic status of surrounding communities*, Washington D.C.: U.S. Government Printing Office.
Vidal, A. (1992) *Rebuilding Community: A National Study of Urban Community Development*, New York: Community Development Research Center.
Vidal, A. and D. Keating (2004) "Community development: current issues and emerging challenges," *Journal of Urban Affairs*, 26, 2: 125–137.
Wortham, S.M. (2006) *Counter Institutions: Jacques Derrida and the Question of the University*, New York: Fordham University Press.
Young, I.M. (1990) *Justice and the Politics of Difference*, Princeton, NJ: Princeton University Press.
—— (2000) *Inclusion and Democracy*, Oxford: Oxford University Press.

11 Fighting for Just Cities in capitalism's periphery[1]

Erminia Maricato
Translated by Bruno G. Lobo and Karina Leitão

INTRODUCTION

Given the (temporary?) defeat of socialist utopias at the end of the twentieth century, is it still possible to achieve more justice in cities in the beginning of the twenty-first century?[2]

Given the territorial mobility of capital in the so-called era of globalization, which in its search for locational advantages condemns cities to abandonment and ruin through unemployment, what are the chances of inverting such processes?

Given that the neoliberal ideas that guide the restructuring of capitalist modes of production impose the deregulation and privatization of public services and undermine the welfare state, how can public agencies meet the needs of those that cannot pay market prices for housing, health care, sanitation, and education?

Does a debate on just cities, as proposed in this book, still make sense considering the following and other features of neoliberal globalization, which are urban in nature or which markedly impact the urban context: unemployment or precarious labor relations, territorial fragmentation, interurban competition, the exacerbation of the "city of the spectacle,"[3] and the rise of the "city-state," the "city-corporation" and the "merchandise city" (Arantes *et al.* 2000)?

If we consider also that those changes have ground down on societies which have not yet ensured universal social rights (such as rights to social security, health, and education) and most of whose economically active population remains in informal activities, what are the odds of fighting the increasing urban inequality in societies that combine a politically subordinated connection to the post-modern condition with the maintenance of pre-modern relations?

Globalization, like Taylorism and Fordism before it, is creating a new man and a new society by transforming states, markets, labor processes, aesthetics, goods, habits, values, culture, social and individual subjectivity, and the production of space and environmental relationships.[4] Although these transformations impact any society or city in the world—regardless of their being labeled winners, losers, or outsiders[5]—this chapter will focus on the context of peripheral capitalism, especially in Brazil. The case study will address a social movement that, swimming against the current and aiming to achieve a right to the city, has managed to achieve several victories at local and national levels. These victories were first achieved at the end of the military dictatorship, when left-wing mayors won office through direct elections (in the 1980s), and have continued to the present with President Lula's second term in office (beginning 2007), during which investments in housing and sanitation were resumed after 25 years of slashing public expenses due to neoliberal structural adjustment policies. The most significant examples of the advances attained during this period have been: (1) the passage in 2001 of the City Statute, a federal law restricting property rights for those properties that do not fulfill their social function, and (2) the creation of the Ministry of the Cities in 2003, which put into practice an ample public participation policy while defining the Ministry's programs to fight urban injustice.

However, in addition to acknowledging the advances achieved thus far, this study also addresses the limitations of recent efforts in Brazil. Such limitations result from two factors: (1) a renewed submissive political and economic role played by Brazil in the context of "new imperialism"[6] (even while playing the role of an emerging country in international relations) and (2) an unchanged archaic and property-based heritage that has also been renewed under the overpowering effect of global forces. This chapter seeks to highlight the fact that this process takes place together with a strong social and participative dynamic integrated by social movements, political parties, NGOs, professional associations, and unions, which generally restrict themselves to fighting for better living conditions and forget utopias of social change, which would be the route truly capable of ensuring a more Just City.

THE "TSUNAMI" OF GLOBALIZATION IN PERIPHERAL COUNTRIES[7]

Harvey (1989) argues that the restructuring of capitalist modes of production initiated in 1973 strongly affected the asymmetrical relationships between countries through dispossession and industrial relocation. To explain the role of predatory control of finance capital and American hegemony in these changes, Harvey (2003) develops the concept of accumulation by dispossession, which complements and completes the theoretical concept of primitive accumulation. Primitive accumulation involves, *inter alia*, the mercantilization and privatization of land, the violent expulsion of peasants, the slave trade, the misappropriation of assets, the increase of national deficits, and the growth of agribusiness. While accumulation by dispossession maintains many of these processes—perhaps in more radical forms—it also involves new ones, including privatization, bio-piracy and the theft of genetic resources, destruction of natural resources, and patenting of transgenic material. Accumulation by dispossession has been accompanied by the breakdown of the classic division of labor. Producers have relocated entire industries to the periphery in search of cheap labor and flexible environmental laws. These peripheral countries have then become exporters of durable goods (Harvey 2003).

Of everything that has already been written on the subject it is only important to highlight one central element: the rise of a neoliberal hegemony through the successful strategy of "capacity building." In Latin America, the policy recipes determined by the Washington Consensus[8] gained importance because this "capacity building" process depended upon the support of multilateral agencies (with the World Bank and the Inter-American Development Bank as two of the most important). These agencies required the implementation of their recipes as conditions for giving loans. Other fundamental agents in the rise of the neoliberal hegemony included well-paid public intellectuals and scholars, especially in economics. The glamour of prestigious foreign universities and intellectuals has been irresistible. With the growing takeover of peripheral countries' institutions by the academic and professional traditions of core capitalist countries, neoliberal policies have created an army of followers around the world, including those in government agencies and central banks.

The Washington Consensus was also assisted in conquering the hearts and minds of these adherents by the tradition in peripheral countries of copying or reproducing policies originating in core countries while considering proposals of indigenous origin extracted from domestic realities and experiences as unworthy. This traditional detachment of ideology (symbols, culture, values, form) from the productive base generates what Roberto Schwarz (2001) describes as "ideas out of place." The detachment between ideology (inspired by core country capitalism) and the urban reality (of peripheral capitalism) is particularly significant in peripheral country urban planning. Planning proposals and urban laws apply only to fragments of the city, while the remainder is beyond state control and does not follow the rule of law. As we will see, the capitalist private market that functions in the legal city excludes the majority of the population.

The inability of the private market to provide an answer to the needs related to housing and urban services is indisputable, especially with regard to cities in capitalism's periphery. However, the recommendation to slash spending on social policies based on the Washington Consensus was strictly complied with and has been one of the decisive factors in the creation of urban conditions since the 1980s. Abstracting away the differences among countries, for more than two decades investments were held back in housing, transportation, and sanitation, while ideas of market efficiency and state-downsizing gained force. However, contrary to what the neoliberal discourse made us believe, states did not become smaller, but were adapted to meet the needs of market growth and financial capital accumulation. As stated by Kurz, market expansion inevitably leads to the growth of the state.[9] Even in core countries social investment expenditures relative to GDP have increased, as revealed by the World Bank (1991). These provide evidence of the ideological and contradictory nature of a neoliberal discourse that attempts to conceal its true intentions: privatizations and deregulations intended to make room for the growth of transnational conglomerates at any cost.[10]

In spite of reiterated statements in the mainstream that growth in formal sector employment, including particularly vast populations in China and India, demonstrates that globalization has been socially inclusive, it is impossible to hide globalization's real effect on cities throughout the world, especially in a context of late urbanization. The tragedy resulting from the implementation of neoliberalism in cities of the peripheral world is well described by Mike Davis in *Planet of Slums* (2006).[11] This chapter illustrates this impact further with data on the Brazilian reality of housing and urban development, which is the basis for the case study below. The data presented herein, interpreted appropriately, may serve also as a general indicator of peripheral capitalism. There are significant differences between cities within the peripheral world and even between cities within Latin America. Brazil is the world's tenth largest economy (Brazilian Institute for Geography and Statistics (IBGE), but it is also the most unequal country in Latin America and the Caribbean and one of the most unequal countries in the world (World Bank 2006). And, obviously, Brazilian cities reflect this fact. However, structural similarities and determinations related to the international division of labor allow the use of generalizations concerning the peripheral world. In regard to Latin America, similarities are also due to shared Iberian colonial roots, as we will later see.

In Brazil, the impact of neoliberal fiscal adjustments began in 1980 and immediately affected economic growth and unemployment rates. Between 1940 and 1970, the country's economy grew at 7 percent annually. During the 1980s and 1990s, growth declined to 1.3 and 2.1 percent annually, respectively. These growth rates were not sufficient to incorporate the young population into the labor market. Throughout the 26 years since the early 1980s, unemployment rates have remained high. According to IBGE, in 2006 two of every three workers were unemployed or underemployed. The impact of unemployment on cities is flabbergasting. The lack of work, particularly for males, combined with a lack of mobility in peripheral neighborhoods and slums (*favelas*), may explain part of the explosive growth in urban violence. While violence was not a central issue for Brazilian cities until the end of the 1970s, it has since become so important that the IBGE claims it started impacting male life expectancy in the 1990s. The annual homicide rate (number of murders per 100,000 inhabitants) increased from 11.7 in 1980 to 27.0 in 2004 (IBGE).

While the Brazilian population grew at an annual rate of 1.9 percent between 1980 and 1991, the population living in slums grew 7.6 percent. In the following decade (1991 to 2000), while the total population grew 1.6 percent, the population living in slums grew 4.2 percent. Slums not only exploded but also, more significantly, strongly densified.[12] The current condition of public transportation reveals the sacrifices the population in the periphery of cities has to make, especially in the major metropolises. Ticket price increases have been confining part of the population to immobility, and the lack of regulation has resulted in increased informal, or illegal, transportation services. According to the Ministry of the Cities and the National Association of Public Transportation (ANTP), the number of people using public transportation has been decreasing, together with population mobility. In 2005, 35 percent of commuting activity in metropolitan areas in Brazil was done on foot, 34 percent by public transportation and only 29 percent by car, despite the latter representing the hegemonic matrix of metropolitan urban mobility (ANTP 2005). Although the increased reach of health care services decreased child mortality between 1940 and 2008, the lack of investment in environmental sanitation has brought epidemics like dengue fever and yellow fever back to Brazilian cities. Both reached alarming proportions in some Brazilian cities in 2007 (yellow fever in the Goiania Region) and 2008 (dengue fever in the metropolitan region of Rio de Janeiro).

The roots of the unjust city: the legacy of patrimonialism

The search for social justice in peripheral countries is hindered by factors other than the recent processes triggered by neoliberal globalization. In the case of Latin American countries that went through Iberian colonization, there are two important conditions derived from this colonial relationship. First, the persistent lack of political autonomy and fragility of internal markets has not allowed for inclusive social and economic development within the framework of capitalism.[13] The relationship between international capital and local élites with complementary interests has for centuries resulted in the exportation of wealth, the appropriation of large rural estates and the hindrance of internal market development.

> Latin America's financial dependence due to foreign debt, from time to time, results in submission to policies adopted by international creditors. This has been taking place ever since the independence of our countries. This happened during the "primary exportation model," and in the second stage during the policy of "import substitution" from the 1950s onward (between the 1930s and the 1950s there was no significant foreign debt, and economic and cultural policies had more autonomy). Foreign debt service, from time to time, becomes such a heavy burden that it leads to crises in the balance of payments, which are generally followed by financial speculation and bankruptcies of banks and companies.
>
> (Tavares 2006: 38)

The image of slum dwellers with cell phones and other electronics (TVs, DVDs) while lacking access to water and sewage systems captures the contradictions produced in countries whose destinies are for the most part decided externally. Market expansion has not only produced consumption objects but also values and desires. Happiness has been linked to a lifestyle. The merchandise fetish reigns absolutely and, through its narcotic power, induces society to ignore basic social needs that have failed to be made priorities. Economist Maria da Conceicao Tavares labels this process of modernization in which elements of premodernism are retained "conservative modernization" (Tavares and Fiori 1993). It consists of a capitalism in which labor relations are characterized predominantly by informality.[14] The analysis of various other authors (Oliveira 2003; Schwarz 2001) suggest that this capitalism subordinates whole societies, including non-capitalist relations, which, as we will see, is evident in cities.

Second, élites are strongly oligarchic and patrimonial. Patrimonialism, or clientelism, has the following characteristics: (1) personal relationships and exchange of political favors are central to public administrations; (2) the public sphere is regarded as something private and personal; and (3) there is a direct relationship between patrimonial property, political power, and economic power. In such contexts, the application of law is unpredictable when dominant interests are at stake. Progressive laws can result in conservative decisions when applied to the interests of the élites and capital, as courts also are influenced by personal relationships.

This deserves a note to explain the stratagems involving law enforcement with regard to property law in Latin America. When breaking the law (i.e., illegal occupation) becomes the norm and the norm (respecting urban laws) becomes the exception, there arises tension and arbitrary power in enforcing the law, both in court decisions and in urban management. The law is enforced according to the specific circumstance and gives differential treatment to different social classes by maintaining the political relations of submission. Therefore, most of the population that lives in illegal conditions remains dependent on personal relationships based on the exchange of favors. The maintenance of political power—or maintenance of dependence and submission—is the cause of this complex pattern of law enforcement.

One characteristic of conservative modernist capitalism is limited private housing markets that serve only a small portion of the population. In core countries, an average of 80 percent of the population has access to private housing markets, while 20 percent is dependent on public subsidies. In peripheral countries the opposite occurs: private markets have limited reach, are socially exclusive and highly speculative. The real estate market is specialized in producing luxury goods, which are typical from an industry that maintains characteristics of manufacturing. It is estimated that in Latin America[15] as a whole, only 20 to 40 percent of the population have access to housing through formal private markets. As public policies cannot address the needs of the remaining 60 to 80 percent, the deprived population is left on its own. Amongst those excluded from housing markets, one can find blue-collar workers, public servants and bankers.

Salaries are not sufficient to guarantee the reproduction of the labor force. This is a central element in the spatial production of peripheral cities:

urbanization with low wages. Housing is not produced by the private legal market, not even with public policy support. Despite recent reports on urban housing conditions throughout the world, there is no rigorous data quantifying such conditions in peripheral countries (UN-HABITAT 2003; ECLAC 2004). Portions of cities with extents that vary from country to country (in some cases including the majority of the population) are built by the tenants themselves without proper technical knowledge (of engineering or architecture), without proper financing, and without taking into consideration property, planning, laws or building codes. Nevertheless, one cannot argue that the state is absent from such processes, as institutional political relationships are sustained through clientelistic practices maintained by political parties, members of congress, and executive administrations.

Urban residents also turn to environmentally sensitive areas, such as banks of streams, rivers and reservoirs, steep slopes, mangroves, flood areas, and valleys, which are subject to environmental legislation that make them worthless to the formal housing market. Essentially, these areas are all that is left to the vast majority of the population. The consequences of such large invasions include, *inter alia*, the pollution of water sources and reservoirs as well as many casualties due to landslides, floods, and epidemics.

One of the main causes of the housing deficit and price inaccessibility is the retention of land and vacant buildings in the search for higher rents. In some Brazilian cities, including Campo Grande, Goiânia, Cuiabá, and Palmas, the vacant areas served by infrastructure could accommodate more

Figure 11.1 São Paulo: photo by Helena Galrão Rios

than twice the population of these cities. On the other hand, in the larger metropolises, especially in Rio de Janeiro and São Paulo, the number of vacant buildings in urban areas is increasing, and is rapidly approaching the housing deficit of both cities, as illustrated in Table 11.1. The empty buildings are located in central areas and therefore fully serviced by urban-infrastructure and services at international standards.

While vacant housing stock languishes in the central areas of Brazilian metropolises, the lower-income suburbs continue to expand horizontally. The expansion is based on a model with serious consequences for a nation with limited resources. While they display a lack of environmental sustainability similar to that of sprawling U.S. suburbs, the peripheries of Brazilian cities (and Latin American cities more generally) are more problematic because they are not subject to the same extensive infrastructure investment and because the household automobile ownership rates are much lower. The dispute over rent from land and urban property is therefore central in promoting urban injustice in Latin America.[16]

TOWARDS URBAN JUSTICE: THE SOCIAL MOVEMENT FOR URBAN REFORM

Moving in a direction opposite from the neoliberal dynamics described above, a social movement called "Urban Reform" was created in Brazil. The movement united social movements (housing, transportation, sanitation), professional associations (architects, lawyers, planners, social workers, engineers), unions, universities and research centers, NGOs, members of the Catholic Church (with the return of the religious movement Liberation Theology), and public servants, as well as mayors and progressive senators. At the time the new Brazilian constitution was being enacted (1987) after the end of the military dictatorship, these social movements came together to form the Forum for Urban Reform with the purpose of providing a common platform for

Table 11.1 Brazilian municipalities with many vacant buildings

Municipality	*Total buildings*	*Total vacant*	*Percent vacant*
São Paulo (SP)	3,554,820	515,030	14.5
Rio de Janeiro (RJ)	2,129,131	266,074	12.5
Salvador (BA)	768,010	98,326	12.8
Belo Horizonte (MG)	735,280	91,983	12.5
Fortaleza (CE)	617,881	81,930	13.3
Brasília (DF)	631,191	72,404	11.5
Curitiba (PR)	542,310	58,880	10.9
Manaus (AM)	386,511	51,988	13.5
Pôrto Alegre (RS)	503,536	46,214	9.2
Guarulhos (SP)	336,440	43,087	12.8

Source: IBGE/Census 2000—Preliminar Sinopis.

Figure 11.2 São Paulo: photo by Isadora Lins

their fragmentary claims, including public participation in land use decisions and planning policies as well as a newly defined right to the city. The right to the city in this case refers to opposing illegal, segregated, distant urban peripheries that lack infrastructure, services, and urban amenities. Housing movements, consistently the majority in the Forum, started focusing on location and reflecting on the necessity of longer term strategies, such as reforming Brazilian property law.

This movement grew along with the general social demand for political freedom. In spite of the arrest of union leaders, important strikes in the industrial regions of ABC[17] in the metropolitan region of São Paulo at the end of the 1970s and the beginning of the 1980s symbolized a strong questioning of the military régime. The Workers' Party (*Partido dos Trabalhadores*, or PT) was created in the 1980s, providing a platform for urban and rural movements and bringing together social movements as well as Catholics and former guerrilla fighters. In the same period, the Central Workers' Union (*Central Unica do Trabalhador*, or CUT) and the Center for Popular Movements (*Central de Movimentos Populares*, or CMP) were also created.

The election of progressive mayors at the beginning of the 1980s initiated the urban reform movement's experience with local government (outside of the state capitals). Programs for urban development and land regularization as well as programs to urbanize peripheral neighborhoods became more participatory, incorporating local residents into public administrations' decision-making processes. This dynamic gained a new energy with the election of left-leaning mayors in state capitals, including São Paulo and Pôrto Alegre. Pôrto Alegre's participatory budgeting may have been the most significant experience to date of social control over local public resources.[18]

A new period in Brazilian politics started in the late 1980s with the end of the dictatorship and the return to direct elections in the capitals (1985) and the presidency (1989). In 1987 a new National Assembly formed to develop a new constitution. It is during the campaign to influence the content of this constitution that urban movements gathered as the Forum for Urban Reform. Two of the most significant achievements were the inclusion of a Popular Initiative Amendment[19] and the inclusion of the social function of property and the social function of the city in the 1988 Brazilian Constitution.

The Popular Initiative Amendment, which was signed by more than 160,000 voters from around the country, was presented to the National Assembly in 1987 by six civil society organizations. For the first time in the history of Brazil, the Federal Constitution included a section dedicated to cities, but its application was dependent on federal regulation. The debate initiated by the National Congress and by the regaining of democratic freedoms allowed for the National Forum for Urban Reform to take place, gathering together the most important urban social movements throughout the country.

Despite constant pressure from the Forum for Urban Reform, the urban section of the Federal Constitution was only regulated 13 years later by the National Congress in the City Statute (*O Estatuto da Cidade*, Federal Law n.10.257/2001). This law provided a legal platform for dealing with the urban question. From a legal point of view the change was profound: the law enacted the social function of property.[20]

It instituted, in accordance with master plans, penalties for unoccupied or underused buildings. It also introduced new public planning instruments and restructured existing instruments that had been fragmented and disconnected. Changes instituted by the new law include: mandatory master plans for municipalities with more than 20,000 inhabitants, compulsory transportation plans for cities with more than 500,000 inhabitants, and mandatory reports on the environmental and neighborhood impacts of large developments. Public participation was also made compulsory in the elaboration of planning documents and policies. There were also juridical instruments for land and housing regularization (since it concerns single properties up to an area of 250 m^2). Through these and other changes, the City Statute has institutionalized the urban sphere in a country where, as previously shown, radical *laissez-faire* capitalism is combined with an exaggerated bureaucracy applied in a discretionary way that is dependent on relations of favor and power.

A long list of the National Movement for Reform's significant victories proceeds from the first experiences in democratic municipalities at the beginning of the 1980s and the subsequent expansion and organization of urban social movements, including land occupation in cities. With the election of

Luiz Inácio Lula da Silva ("Lula") in 2002, the Ministry of Cities was created to respond to the demands of the social movements. The creation of this new ministry also symbolizes the advent of a new phase of institutionalizing social demands specific to the country's urban conditions. A summary of the most important social advances since 1987 is found in Table 11.2.

The creation of the Ministry of Cities and the opening up of various councils to public participation were the determining factors in guaranteeing the main advances listed in Table 11.2 after 2003. From its inception, the Ministry was staffed by a team of professionals with unique backgrounds:

Table 11.2 Summary of events since 1997

1987	Constitutional Amendment by popular initiative subscribed to by six entities of civil society; Creation of the National Forum for Urban Reform formed by entities of civil society.
1988	Enactment of the Federal Constitution with two sections focused on urban issues, a first in Brazilian history.
1991	Introduction of a bill on a national fund of public housing as an initiative of civil society subscribed to by one million voters (passed into law in the Federal Chamber as National Fund of Housing of Social Interest in 2005).
2001	Enactment of "The city statute," a federal law which amends the Federal Constitution of 1988 to specifically uphold the social function of property.
2003	Creation of the Ministry of Cities; National Conference of Cities held—originated from a process involving 3400 municipalities in all states of the Brazilian federation. More than 2500 elected delegates attended the conference in order to debate the "National Policy of Urban Development" (other conferences took place between 2005 and 2007).
2004	Creation of a National Council of Cities to consult the Ministry of Cities; Creation of the National Program of Regularization of Urban Land Property.
2005	Approval of the Federal law of the National Fund of Housing of Social Interest—created a fund for connecting council activities with social participation, which conditioned the transfer of federal resources to local and state Councils for the creation of Housing plans and other programs. The National Campaign for Participative Master Plans was launched in the same year calling for the elaboration of such plans for all Brazilian cities with more than 20,000 inhabitants.
2007	Lula's administration launched the Program for the Acceleration of Growth (PAC, or "*Programa de Aceleração do Crescimento*") resuming investments in housing and sanitation abandoned for 25 years. The plan of Keynesian influence proposed a series of public works aimed at renovating part of the infrastructure focused on production (ports, railways, highways, electrical plants) as well as social and housing infrastructure. Between 2007 and 2010 R$106 billion will be spent in housing and R$40 billion will be spent in housing and sanitation.

the majority had academic positions; almost all were active militants in left-wing parties (the majority from the PT) or from social movements and labor unions; and the vast majority had previously held jobs in innovative local and regional administrations.[21] Lula chose as minister Olivio Dutra, the former governor of the state of Rio Grande do Sul and former mayor of the city of Pôrto Alegre, where the experience with participatory budgeting was acclaimed and where the first meetings of the World Social Forum took place.[22] However in 2005, two-and-a-half years after its creation, the Ministry of Cities changed direction. Internal disputes in the PT and alliances aimed at guaranteeing majorities in the National Congress led the government to award the Ministry of Cities to a party with a conservative and clientelistic profile. Some of the previous team left the federal government and some continued trying to implement the programs already defined. But the change interrupted the transformative momentum that had originally defined the public agency as a builder of federal policies on urban development.[23]

THE LIMITS OF PARTICIPATORY POLITICS: PROBLEMS PERSIST

In spite of advances in fighting poverty and promoting social policies in sectors of the administration retained by the left, one is compelled to recognize that the impact on the broader economy and politics has been limited. The neoliberal model persists with little change.

The Program for Acceleration of Growth (*Programa de Aceleração do Crescimento*, or PAC), a development of Lula's second term, proposes a set of public works that follow a Keynesian plan. Though the plan revives investments in sectors that have been paralyzed since the beginning of the 1980s, like social and economic infrastructure (sanitation, energy, logistics) and housing, it is necessary to highlight the continuing dominance of finance capital. One example of this overwhelming dominance is the draining of public resources into the financial system through interest payments on public debt (both internal and external). The Brazilian Central Bank, whose president, appointed by Lula, was previously Boston Bank's president, set interest rates in Brazil among the world's highest over the majority of recent years. As a result, the Brazilian federal budget for 2007 dedicated 30.59 percent (R$237 billion) of all public resources to make interest and principal payments on public debt. Such allocations of public resources constrain public spending. The same budget allocated only R$40 billion for health care, R$20 billion for education, and R$18 billion for the Family Scholarship program, the country's most comprehensive social program. Considering the impact that social policies have on income distribution and internal market strengthening, one wonders how many lives could be improved if these resources were addressing social needs. Instead, the majority of these resources end up in the hands of national and international banks.[24]

Because of the low capacity for investment, the PAC allocates R$106 billion to housing over four years (2007–2010). Most of the funds are projected to be used in the private housing market because they come primarily from the private and quasi-private sectors. Only R$10.1 billion of the R$106 billion come from the Total Budget of the Union (*Orcamento Geral da Uniao*, or OGU). Even considering the subsidies from other sources, such as the Time of Service Security Fund (FGTS), there is a lack of public subsidy for housing given that 84 percent of the housing deficit is concentrated in households with income levels below three times the minimum wage (approximately US$645 per month in 2007).[25]

The ambiguity of government policies, which arises from the combination of maintaining the interests of finance capital while fighting poverty, is also present in the agrarian reform program. The government provides moderate support for agrarian reform: the fact that it does not use the police to repress land occupations already represents a difference to some militants. However, at the same time, it fails to address the interests of the agricultural industry and rural landowners. Under globalization, Brazil has become an exporter of commodities, including celullose, grains, meat, ethanol, and mineral products. This role has combined with the need to address financial commitments (debt service) to strongly impact Brazilian territory. Consequences of the capitalist expansion sponsored by large corporations and banks involved in agribusiness include the deforestation of the Amazon Region and the conversion of large, unproductive rural properties into international corporate properties. Despite the government's intention, it is impossible to combine policies of family agriculture and programs of land distribution for members of the Landless Movement (*Movimento dos Sem-Terra*, or MST) without confronting the interests of agribusiness and rural landowners, whose properties are in most cases of questionable origin.

In politics, two issues are particularly worth noticing. The first concerns the formation of coalitions to achieve governability, a universal phenomenon that is not just typical of Brazil or other peripheral countries. When ruling progressive parties form alliances with conservative parties and supporters of clientelism, they strengthen the premodern aspects of institutional relations. By trying to aggregate "buyable" votes into a political base, the government becomes hostage to a trade off in which the cost of having legislative support is loss of control over government appointments and parts of the federal budget. Public investments come to be decided according to the logic of financing electoral campaigns.

The PT has followed a similar trajectory. As it became more focused on electoral campaigns during the 1990s, the party increasingly moved away from the social movements and intellectuals that were at its origins. The logic of electoral contest demanded that the party's leaders build an internal bureaucracy disconnected from the party base and that they adopt a hitherto nonexistent pragmatism in order to secure sources of financing capable of meeting the challenges and interests at stake. This change became explicit in 2005, when top members of the party were accused of using illegal methods to buy votes in the legislature. This reorientation of the PT has left a void in Brazilian politics that is slowly leading to a restructuring of political organizations on the left, but it is still too early to predict how the restructuring will play out.

The second notable political issue directly relates to the theme of this paper: urban social movements in their struggle for more just cities. The "participation fever" that has swept public agencies, NGOs, political parties, and social movements is present not only in Brazil. From the World Bank to Via Campesina (International Peasant Movement), the word of the day is participation. As the Brazilian philosopher Paulo Arantes states: "The political sphere has never been so full and at the same time so empty."[26] While it is true that such movements are concerned with important issues such as gender, race, environmental sustainability, sanitation, and housing, they are simultaneously fragmented and divided, treating parts as the whole and ignoring concerns for the future of society. Such movements have abandoned the search for alternative social change, despite the fact that there is no future for society if we sustain the same production and consumption patterns of the last 150 years.

Lula's government has promoted 40 national, municipally organized conferences focused on issues including youth, racial equality, elderly rights, cultural policies, women's rights, disabled rights, and rights for children and teens. Since 2003, approximately two million people have attended these conferences. On average, 1500 delegates participated in each of the three National Conferences on Cities (2003, 2005, 2007). These were preceded by local and state conferences focused on discussing a foundation for building public policies and fighting for more just cities. The conferences elected the City Council, a consultative body composed of representatives of different social groups that advises the Minister of Cities. There is no doubt that this process contributed to the expansion of the debate on cities and the addition of members to existing social movements. However, the premodern forces previously discussed affect the implementation of participation in governmental spaces.

The country's main urban movements voluntarily constrain themselves within the institutional and governmental agenda. Some have evolved into an oligarchic model with the same leaders over time who enjoy personal control over local bases and close connections with specific members of Congress. This reproduces the phenomenon of electoral obligation, and in some cases, their own members run for election. Strategies of personal survival—reinforced by high unemployment rates—also form part of the relationship between leaders and political bases. Instead of increasing democracy and subordinating the state, the struggle to amplify the control of social movements over the government is led by the particularistic claims of their bases and not with the objective of building policies based on universal rights. One cannot say that this situation is prevalent at the federal level. There is, nevertheless, a strong movement that seeks to change the regulations that originally led to such participatory processes.[27]

Of all the constraints on the struggle for a more Just City that have been discussed, the most important is the resistance to the application of the City Statute. The law interferes with interests that form an essential part of the Brazilian society, as previously mentioned. Real property has always been connected to political and economic power. And as discussed earlier, the rule of law is also subject to power relations. While this is true throughout the world, in societies as unequal as Brazil, local circumstances are even more important in the application of laws. At the same time, the City Statute is an inherently difficult law to implement. The instruments related to the social role of property and the public capture of increases in property values are subordinated to the City Master Plan, as mentioned above. That implies that City Councils have to make decisions on the implementation of public instruments aimed at limiting the increase of property prices in urban areas. Deputy mayors, however, tend to be close to landowners and developers, and local authorities have a tradition of personal and family appropriation of rents from land and real estate.[28] For this reason, the City Statute has to date had a larger impact on the discourse of planners and lawyers than on policies of urban inclusion. The excitement the City Statute engendered when it was passed captivated many social movements that had been prioritizing the fight in the legal arena. However, following passage, the élite used multiple strategies, like the one described next, to stop or to at least postpone the enforcement of the "social function of property" in Brazilian cities.

The 1988 Constitution was vague when addressing the enforcement of the social function of property, and enforcement was postponed until a specific law was passed to regulate it: the City Statute, which was enacted 13 years after the 1988 Constitution. In turn, the City Statute passed in 2000 displaced the enforcement of its mechanisms onto master plans, in compliance with the Constitution. Most of the master plans drafted after 2001 resulted in vague and general texts that deferred the enforcement of the City Statute's mechanisms to municipal laws. However, to this date, these municipal laws have rarely been passed.

The law is the result of a long struggle and serves as a reference for social movements that have put many of their hopes for achieving the right to housing and the right to the city into the law's success. This is illustrated by the current mobilizations in several Brazilian cities supporting Master Plans that, through public participation, include instruments defined in the City Statute. In São Paulo, social movements constituted in 2006 a Front for the Support of the Master Plan (*Frente em Defesa do Plano Director*) enacted in 2004, in reaction to the mayor and city council's desire to exclude the advances of 2004 in addressing the interests of the real estate industry. After winning the dispute for the Brazilian Constitution in 1988, after the enactment of the City Statute in 2001, and also after the development of participatory master plans in 2004, the implementation of the social function of property is about to begin.

In conclusion, the enforcement of the legal concept of the social function of property, which is central to guaranteeing social justice in Brazilian cities, has apparently advanced with the enactment of the City Statute in 2001. Although social movements—and also urban planners and lawyers—have celebrated the achievement of a legal framework that restrains property rights, the resistance to its implementation shows that the fight has barely started.

The creation of the Ministry of Cities in 2003 has also been celebrated by social movements as a great victory. However, its command was transferred from a left-wing to a right-wing orientation in 2005. Given all the considerations made at the beginning of this text, we can state that it is impossible to separate the fight for more just cities in Brazil from the more general fight against subordination to the interests of financial capitalism and patrimonial interests, which are being renewed with globalization.

NOTES

1 Thanks to Cuz Potter for his careful editing and review of this chapter.
2 See the pessimistic thoughts of Perry Anderson (2000, 2007).
3 Referring to Guy Debord's concept of the "Society of the Spectacle" (Debord 1992), we could say that the "city of the spectacle" is dominated by the image, the appearance. It is the city of the monologue, of passivity, of the true alienation factory.
4 On the impacts of Fordism on man and society, see Gramsci (1949). On the concept and impacts of globalization or "*mondialization*," see, among others, Anderson (2000, 2007), Harvey (1989), Stiglitz (2002), Chesnais (1994), Arrighi (2007), Bauman (2003), and the International Seminar Annals of the Ibero-American Network of Researchers on Globalization and Territory (*Red Iberoamericana de Invetigadores sobre globalización y territorio*) (Available HTTP: http://www.uaemex.mx/pwww/rii/home.html).
5 This statement was made by Peter Marcuse during a talk given in the Graduate Program at the School of Architecture at the University of São Paulo in May 1998. In fact, as we will see, productive restructuring also impacts spaces which apparently have not yet been reached by capitalist relationships.
6 This statement is derived from Harvey (2003).
7 Arrighi (1997) proposes the classification of peripheral and semi-peripheral countries. In this article, no distinction is made between the periphery and semi-periphery of capitalism. Other categories can be found in "mainstream" literature (e.g., low- and middle-income countries; less-developed regions and least-developed countries; emerging and poor countries). One cannot ignore that globalization promotes simultaneously, at the planetary scale, the deepening of interdependency, differentiation, and homogenization. Nevertheless, the core–periphery approach remains consistent and offers us an efficient methodological resource, especially when considering the cities.
8 The Washington Consensus policy recommendations are described in the report *Latin American Adjustment: How Much Has Happened?* by John Williamson (1991, Institute for International Economics, Washington) and presented at a meeting in Washington in 1989. Years later the Washington Consensus report was complemented by a follow-up to guide its army of followers. See John Williamson (1994) "The political economy of policy reform," Institute for International Economics, Washington.

9 "[T]he more market economy increased structurally, by encompassing the entire social production system and by becoming the universal way of life, the more the activities of the State needed to be increased. We are, therefore, face to face with an unequivocally reciprocal relationship" (Kurz 1997: 96).
10 The aggressiveness and the disregard for ethics by some transnational corporations can be verified in the documentary movies "The Corporation" (2004, directed by J. Abbott, M. Achbar, and J. Bakan) and "Le monde selon Monsanto," whose director has written a book version (Robin 2008).
11 The quantitative data used by Mike Davis should have been more rigorous. However, one cannot disagree with the general orientation of the critiques present in the text. This comment is detailed in this author's postscript to the Portuguese edition of Mike Davis' book.
12 According to Jordan and Simioni (2003), the housing deficit in Latin America and the Caribbean went from 38 million in 1990 to 45 million in 2000.
13 See, among others, the work of Caio Prado Jr., Celso Furtado (1995), Florestan Fernandes (1975), and Arif Hassan (2008).
14 Francisco de Oliveira (2003) puts in check the definition of informality in the productive restructuring context, observing that informal workers are a capitalist workforce and that the goods and services provided by large corporations, such as those in the telecommunications industry, incorporate the low-income population into the consumer market. According to the author, no one is excluded.
15 In more recent years Chile (1990s), Mexico, and Brazil (after 2000) have been experiencing a real estate boom, whose impact on production and on the consumer market still await more conclusive analyses.
16 An extensive bibliography on urban land in Latin America can be found in the publications and website of the Lincoln Institute of Land Policy.
17 The ABC Region is constituted by the municipalities of Santo André, São Bernardo, São Caetano, and Diadema, among others. It is located in the Metropolitan Region of São Paulo, an area where Fordist industry was installed in the 1950s and where strong workers' unions changed the political history of the country in the 1980s. This region has been impacted by the process of restructuring capitalist production in the 1990s.
18 The members of the Participatory Budgeting Council were chosen in direct elections in order to debate and decide City Hall's investment priorities. The City Council, which is legally responsible for approving the Municipal Budget, were constrained to accept the decisions of this kind of "Popular Assembly."
19 The 1988 Brazilian Constitution established the possibility of introducing bills through public initiative to the National Congress if they are signed by 1 percent of the nation's voters.
20 According to the second paragraph of article 182 of the Brazilian Constitution: "Urban property achieves its social function when it meets the essential city organization requirements set forth in the Master Plan."
21 The financial resources for PAC Housing come from the private market (Brazilian System of Savings and Loans, or SBPE) or private savings (39 percent), a semi-public fund, an employment fund created through salary contributions (35 percent), funding from states and municipalities (17 percent) and the federal budget (9 percent). Source: www.brasil.gov.br/pac.
22 The author of this paper was invited by Lula to be part of the transition team that created various administrative units, including the Ministry of Cities. Afterward, the author stayed on in the administration as Deputy Minister until 2005.
23 Besides the urban issues mentioned, it is worth recognizing that some of the social policies developed by Lula's government are impacting, though lightly, the income distribution in Brazil. The Family Scholarship Program (*Bolsa*

de Familia) was intended to guarantee a minimum allowance to the poorest 40 percent of the population, and was able to assist 11 million families (about 25 percent) in 2007. Between 2002 and 2007 about 20 million people left income categories E and D (E = US$149.60 and D = US$285.10 average monthly income in 2007) and started to be classified as C (C = US$552.70 average monthly income in 2007), based on the categories defined by the "Brazilian Criterion," according to the Brazilian Association of Investigation Companies and the Brazilian Institute of Geography and Statistics. Between 2003 and 2007, 9.7 million Brazilians escaped poverty. The minimum wage (approximately US$215 in 2007) saw a real increase of 32 percent in the same period (Federal Government 2008). (These figures reflect the exchange rate on 31 December 2007, when one Real corresponded to USD 0.5646.)

24 Between 1978 and 2007, public debt quintupled despite payments of USD $262 billion more than the value of all loans (Rede Jubileu Sul/Brazil, 2nd edn, 2008). Part of the interest payments were due to Brazil's risk rating, which is defined by international agencies that penalize poor, indebted countries but failed to recognize the risk of the bubble in the US housing market.

25 Although the subsidies are fairly insignificant, the federal government is implementing the largest slum upgrading program in its history. Ultimately, this is unlikely to have a significant impact in the total number of slum inhabitants.

26 This statement was made by Paulo Arantes during a talk given in the Post-Graduate Program at the School of Architecture at the University of São Paulo in August 2008.

27 It is possible to note the beginning of new types of urban movements that have begun without being swallowed by the institutional space. The majority of social movements that have opposed such trends are rural and not urban. In spite of the difficulties, the Landless Movement (MST) continues to make the effort, claiming immediate needs and arguing over a strategic project that takes into account cultural, environmental, ideological, economic, and political issues. The movement is careful to renew its leaders and organization, democratically discusses each step, and above all maintains its independence despite support from public resources and international donations. The movement's emphasis on education, contrary to common sense, reveals the essential role occupied by education and communication. This is why the movement has been criminalized by the conservative media and judicial system. Between 2000 and 2005, 223 peasants, union members, lawyers, and religious people connected to the movement were murdered.

28 These statements are based on the experience of the author as Secretary of Housing and Urban Development in the City of São Paulo in the first PT administration of the city (1989–1992), as well as the provision of consulting services to several Brazilian and foreign cities.

References

Anderson, P. (2000) Renewals, *New Left Review*, Jan.–Feb., 1: 5–24.

—— (2007) Jottings on the Conjuncture, *New Left Review*, 48: 5–37.

Arantes, O., Vainer, C., and Maricato, E. (2000) *A cidade do pensamento único*, Petrópolis: Vozes.

Arrighi, G. (1997) *A ilusão do desenvolvimento*, Petrópolis: Vozes (translated from the original: "Workers of the world at century's end").

—— (2007) *Adam Smith in Beijing*, New York: Verso Books.

Associação Nacional Transporte Público (ANTP) (2005) *Transporte Metroferroviário no Brasil: Situação e Perspectivas*. Available HTTP: http://www.antp.org.br.

Bauman, Z. (2003) *Liquid Love: On the Frailty of Human Bonds*, Malden, MA: Polity Press.
Chang, H. (2002) *Kicking Away the Ladder: Development Strategy in Historical Perspective*, London: Anthem Press.
Chesnais, F. (1994) *La mondialization du capital*, Paris: Syros.
Davis, M. (2006) *Planet of Slums*, New York: WW Norton.
Debord, G. (1992) *La société du spetacle*, Paris: Gallimard.
Economic Commission for Latin America and the Caribbean (ECLAC) (2004) *The Millennium Development Goals: A Latin America and Caribbean Perspective*, Santiago: ECLAC.
Fernandes, F. (1975) *Capitalismo dependente e classes sociais na América Latina*, Rio de Janeiro: Zahar.
Fiori, J.L. (1995) *Em busca do dissenso perdido*, São Paulo: Insight.
—— (1997) *Os moedeiros falsos*, Petrópolis: Vozes.
Furtado, C. (1995) *Formação Econômica do Brasil*, São Paulo: Cia.Editora Nacional.
Global Urban Observatory (2003) *Slums of the World: The Face of Urban Poverty in the New Millennium?*, New York: UN-HABITAT.
Governo Federal do Brasil (2008) "Relatório aponta redução da pobreza e desigualdade no país," available HTTP: <http://www.agenciabrasil.gov.br/noticias/2007/08/29/materia.2007–0829.7514974886/view>.
Gramsci, A. (1949) *Americanismo e Fordismo*, Milano: Universale Economica.
Harvey, D. (1989) *The Condition of Postmodernity: An Enquiry into the Origins of Cultural Change*, Oxford: Blackwell Publishers.
—— (2003) *The New Imperialism*, New York: Oxford University Press.
Hassan, A. (2008) "Global capital and the cities of the south: with investments in the picture, money will shape the souls of cities," *Cluster*, 7: 126–131.
Instituto Brasileiro de Geografia e Estatística (IBGE) Statistical database. Available HTTP: http://www.ibge.gov.br.
Jordan, R. and Simioni, D. (2003) *Gestion urbana para el desarollo sostenible em America Latina y Caribe*, Santiago de Chile: CEPAL.
Kurz, R. (1997) *Os últimos combates*, Petrópolis: Vozes. (This book is a Portuguese version of various articles written by the German philosopher, co-founder, and editor of *EXIT!* Magazine, Kritik und Krise der Warengellschaft.)
Marcuse, P. (1997) "The enclave, the citadel and the ghetto: What has changed in the post-Fordist US city," *Urban Affairs Review*, 33(2): 228–264.
Maricato, E. (1996) *Habitação e cidade*, 7th edn, São Paulo: Saraiva.
—— (1996) *Metrópole na periferia do capitalismo: desigualdade, ilegalidade e violência*, São Paulo: Hucitec.
—— (2007) "As idéias fora do lugar e o lugar fora das idéias," in O. Arantes (ed.) *O e outros: A cidade do pensamento único*, 4th edn, Petrópolis, Vozes.
Oliveira, F. (2003) *Crítica à razão dualista/ornitorrinco*, São Paulo: Boitempo.
Rede Jubileu Sul/Brasil (2008) "ABC da dívida: sabe quanto você está pagando?" available HTTP: http://www.jubileubrasil.org.br e www.divida–auditoriacidada.org.br.
Robin, M. (2008) *Le monde selon Monsanto*, Paris: La découvert (book and video).
Schwarz, R. (2001) *Cultura e política*, São Paulo: Paz e Terra.
Smolka, M. (2005) "El funcionamiento de los mercados de suelo en America Latina," in J.L. Basuado (ed.) *Manejo del suelo urbano*, Seminário Internacional, Corrientes, Argentina: LILP e Instituto de Vivienda de Corrientes.

Stiglitz, J. (2002) *Globalization and Its Discontents*, New York: W.W. Norton.
Tavares, M.C. (2006) "Conferência introdutória," in *Cadernos do desenvolvimento*, Rio de Janeiro: Centro Internacional Celso Furtado de Políticas públicas, Ano 1, n. 1: 37–58.
Tavares, M.C. and Fiori, J.L. (1993) *Desajuste global e modernização conservadora*. São Paulo: Paz e Terra.
UN–HABITAT (2003) *The Challenge of Slums: Global Report on Human Settlements*, London: Earthscan Publications Ltd.
Van Kempen, R. and Marcuse, P. (1997) "A new spatial order in cities?," *American Behavioral Scientist*, 41 (3): 285–298.
Williamson, J. (1991) *Latin American Adjustment: How Much Has Happened?* Washington: Institute for International Economics.
—— (1994) *The Political Economy of Policy Reform*, Washington: Institute for International Economics.
World Bank (1991) *Annual Report 1991*, Online Available: http://go.worldbankorg/IH0OSLPQ0.
—— (2006) *The World Bank Annual Report 2006*. Online. Available HTTP http://go.worldbank.org/KQ3OFEED90.

12 Race in New Orleans since Katrina[1]

J. Phillip Thompson

On August 29, 2005, Hurricane Katrina struck southeastern Louisiana with winds of 100–140 mph and heavy rain. Storm surges left most of the city of New Orleans flooded, some parts beneath 20 feet of water—1300 people lost their lives, 380,000 people were displaced from the city, 105,000 (of 188,000) housing units in the city suffered severe damage. The Bring New Orleans Back Commission, formed by the Mayor soon after the storm, marked many formerly vibrant neighborhoods as "areas for future parkland." The reasons for the flood in New Orleans go deep into the city's history. They are tied to the hubris of traditional élites against nature (the mighty Mississippi River), and their haughtiness fed by an even older tradition of human exploitation (slavery) that was a prime source of the poverty and vulnerability to the flood gripping large numbers of the city's residents. If we are to imagine a just New Orleans, this history must be considered and social reconstruction must precede physical reconstruction.

During the midst of Mayor Ray Nagin's reelection campaign in the Spring of 2006, just months after the flood, Bill Rousselle, a longtime black political activist in New Orleans, commented during a lunch conversation with a few nationally known black labor leaders that the flood presented the "biggest opportunity we've had since Reconstruction" (Rouselle 2006). What he meant, and what hardly needed explanation at this lunch table, was that the causes for the disaster in New Orleans, and problems the people of New Orleans must still overcome, lay deep in the region's slave past. Enslavement of Africans and some Native Americans flourished in the early 1700s in Louisiana and lasted until 1865. For 11 years, Reconstruction governments ruled Louisiana with the assistance of Northern Union troops. This was the brief period of opportunity Rouselle referenced. Reconstruction was crushed in Louisiana and was followed by nearly 100 years of legal apartheid. During this entire 300-year period, few blacks ever earned a "living wage," lived in a safe and sanitary house, or had a decent education. The civil rights and black power movements offered glimmers of hope for improved circumstances, but not so much in Louisiana, where white racism was even harder to overcome than in Mississippi next door.

Slavery and segregation did not only structure black life in New Orleans; it impacted whites just as much. Despite a radical populist history of their own, poor whites supported rich Louisianans in their shameless exploitation of the black community. Even so, poor whites in Louisiana, who are a larger share of the white population than most places in the nation, were themselves barely citizens. Their "white privilege" enabled them to avoid becoming slaves and offered them the dignity of walking into public libraries, but it did not ensure that they could read, earn living wages, or afford healthcare. Still, poor whites lived in fear of the even worse depravation heaped upon blacks. That is why, in the early 1960s, when Federal courts mandated public schools in New Orleans to finally desegregate—despite over 60 state legislative bills designed to block school desegregation—whites fled the city in droves. New Orleans was the site of one of the most contested school desegregation battles in the country (Fairclough 1995). After desegregation New Orleans experienced steady "white flight."

The city lost about 250,000 whites during the 1960s, and another 150,000 during the 1970s. The city's black population grew only by about 60,000 since 1960. As a result, the race conflict was spatially and politically transformed, and coded, into tensions between "the city" and "the state" (read: "black city" in "white state"). This largely explains why New Orleans was so vulnerable prior to Katrina, and why rebuilding is so difficult post-Katrina. Concentrated in the surrounding suburbs, resentful of the black community's assertiveness and of federal programs targeted to help the black poor, the white majority elected state officials determined to starve the mostly black city. During 25 years of black rule in New Orleans, black mayors' "blackness" was determined by their willingness to fight off white élites attempting to capture the city's institutions (through privatization, receivership, or state control). This is the background of poverty and vulnerability in New Orleans.

The enslavement of 5,000 Africans in the city (and a thriving slave importation industry at the port), the expulsion and extermination of indigenous peoples, the command of a fearful poor white laboring class, the construction of monumental mansions on St. Charles Avenue in New Orleans, the use of poor black and white women as sex-playthings in New Orleans's still

Table 12.1 Population growth by race in New Orleans

Year	*Total*	*White*	*Black*	*White %*
1950	659,000	464,504	195,000	0.70
1960	845,237	579,662	264,000	0.68
1970	593,471	323,420	267,308	0.54
1980	496,000	173,554	307,728	0.34
1990	484,000	136,000	326,000	0.28

Source: U.S. Bureau of the Census, Census of Population and Housing.

vibrant sex industry: all of this could not occur without a high level of wealthy white, male, hubris. With so many accomplishments under their belts, it is no wonder that they sought to subdue, direct, and humiliate the mighty Mississippi River. With a low regard for any power above their own, they did not take seriously the role of the barrier islands as a safeguard against hurricanes passing through the Gulf of Mexico, and they built a city on a river destined to move farther west (Foster and Giegengack 2006). Social subjugation, racism, and dominance of the river were parts of a whole.

Before the storm

Following the 1994 Los Angeles earthquake, HUD Secretary Henry Cisneros said that the earthquake "bares the truth." He was referring to the pre-existing poverty, sub-standard housing, and *de facto* racial segregation in Los Angeles. Disasters usually hit the poor the hardest, excavating hidden pre-disaster tensions. This was certainly true of New Orleans. New Orleans was in crisis long before Katrina, making recovery slow and uncertain. Government did little to help the poor and vulnerable before the storm, and did little after.

The state of Louisiana did not have a housing development agency prior to the storm to support affordable housing development across the state. After Katrina, programs and procedures had to be created and negotiated, staff had to be hired and trained wholesale. This process is still under way. Community development groups in New Orleans were under-funded and under-staffed before the storm. The most advanced community development non-profit in the city produced less than 20 units of housing per year prior to Katrina. There is currently a need for tens of thousands of affordable housing units and little civic capacity to do it.

There was an educational crisis in New Orleans prior to the storm. New Orleans had a 40 percent literacy rate (Dyson 2006). Mayor Nagin, speaking three months before the storm, described schools in the city as having already been struck by "a category five hurricane." Over 50 percent of black ninth graders were not expected to graduate high school in four years. The AFL-CIO Building Trades Council, which operates a workforce training center in New Orleans, says that a majority of prospective workers (young adults) entering its center have a 3rd to 4th grade reading and math level (Graham 2008). Construction trades, such as the carpenters and the electricians, require competency at a 12th grade level to begin their apprenticeship programs.

There was an employment crisis pre-storm, and it is worse now. Before the storm, 84 percent of those categorized as poor (earning less than $16,000/year with a family of three) in New Orleans were black. Nearly 40 percent of all blacks in New Orleans were poor pre-storm. Thirty-five percent did not own a car. Many service workers, representing a quarter of all workers before the storm, earned $8.30 per hour, not enough to support a family (Lerman *et al.* 2006).

There was a serious crime problem, mostly related to drug dealing, prior to the storm. New Orleans's murder rate was the highest in the nation in the mid-1990s. The year before the storm, for an experiment, the police fired 700 blank rounds in a single afternoon in a city neighborhood. Fearing retaliation, no one called to report gunfire (MSNBC 2005). Crime is back, clearly related to education and employment problems. Young people with jobs and decent incomes do not often murder other young people for prime drug-dealing corners.

There was an environmental and health crisis in New Orleans prior to the storm. The city is vulnerable to strong storm surges from the Gulf because barrier islands have been decimated to carve out shipping lanes for energy companies. New Orleans is also part of "cancer alley," because of the location nearby of so many refineries and a long history of toxic run-offs (SATYA 2005). The storm brought these problems out in the open, but the underlying problems have been there for a long time.

Why are the local institutions so weak? As mentioned earlier, the political structure in Louisiana, where whites are a majority in the state and blacks in the city, was more deeply split than the river levees during the storm. Moreover, the very idea of government has been under Republican assault in the nation for more than a quarter century. Starved of federal resources, local governments in many states have atrophied. Predominantly black cities like New Orleans have been fiscally starved in addition by hostile white suburban-dominated legislatures. New Orleans began 2006, the year after the storm, with a budget shortfall of approximately $168 million, and with $156 million less than what the city maintained that it needed to operate. The city relied on a $3.5 million grant from a private foundation to organize a rebuilding plan. While government spent hundreds of billions on a "nation-building" project in Iraq, it was missing in action in New Orleans.

Mayoral elections in New Orleans are also racially polarized. New Orleans' first black mayor, Dutch Morial, was elected in 1972. He was a radical advocate who tried to democratize city government and direct attention to the poor. Morial got 97 percent of the black vote and 20 percent of the white vote (showing a significant liberal white trend in the city). New Orleans' second black mayor, Sidney Barthalemy, was viewed as an accomodationist to the white power élite and won office with only one quarter of the black vote and 85 percent of the white vote. Marc Morial, Dutch's son, succeeded Barthalemy in 1994. Marc Morial served two terms and focused on cleaning up the city's notoriously brutal and corrupt police department. Morial also tried to enact a commuter tax for suburbanites working in New Orleans. Like his father, he got the lion share of the black vote and little of the white vote. Ray Nagin, in his first race for public office, was viewed as an accomodationist similar to Barthalemy and lost the black vote to Morial's police chief, Richard Pennington (who was supported by Morial). Nagin won most of the white vote and the election. Following the storm, some core white supporters advised Nagin that they no longer needed him now that the racial balance in the city had shifted toward whites.[2] Nagin then turned to the black community for support, and pledged to make the "right of return" for poor blacks a key policy objective for his administration. This was the context for Nagin's infamous statement in a black church that New Orleans would again become a "Chocolate City." Running as a champion of poor blacks displaced from the city, Nagin won with the lion's share of the black vote. He lost the white vote heavily.

Racial attitudes in the city

A survey of more than 1200 Katrina survivors by Michael Dawson and Melissa Harris-Lacewell showed deep divisions in opinion between black and white survivors of the storm. Nearly all African-American respondents, and a few whites, believed that either government intentionally blew apart a levee in the 9th Ward to prevent an overflow of water into the wealthier white parts of the city, or that corruption in local government was responsible for inadequate levees in poor black areas of the city. The distrust black New Orleanians hold towards government may sound extreme or ungrounded, but familiarity with the city and its history makes their beliefs plausible. During a similar flood in 1927, local officials dynamited levees in a poor area (populated largely by Cajun trappers) of the city to prevent flooding of wealthy neighborhoods and businesses (Barry 1997). This was never forgotten. The treatment of survivors after the storm reinforced their worst fears. Few will forget being stranded on rooftops for days on end, or crowded into the unsafe, unsanitary Superdome. When hundreds of victims sought safety on the other side of the Crescent City Connection Bridge in the neighboring suburb of Gretna, police officers fired shots over their heads and then pointed weapons directly at the crowd. Two paramedics from San Francisco trapped during the storm were among the crowd. They wrote that the police from Gretna "told us that there would be no Superdomes in their city . . . These were code words that if you are poor and black, you are not crossing the Mississippi River—and you weren't getting out of New Orleans" (Johnson 2005). Even six months after the storm there were powerful signals of government neglect around mounds of rubble in what had been a Lower 9th Ward neighborhood. Walking through the mounds of rubble were lay ministers from Billy Graham Ministries. The ministers approached families huddled in front of demolished houses, and they gathered clusters of police officers, leading them in prayer. I asked the ministers why so many people were praying, and they said that there were an unknown number of bodies (perhaps hundreds) still buried under the rubble. The mounds of rubble were sacred ground. The bodies had not been recovered, they said, because FEMA had not released overtime funds for the Fire Department to bring dogs and equipment. I lived in New York during the terrorist attack in September 2001, and remembered clearly that no expense had been spared to recover bodies and identify remains from the World Trade Center. I was stunned at the contrast in New Orleans.

If New York is the standard, then residents of the 9th Ward were not being treated as U.S. citizens. Incredulously, six months after the disaster, families still had not been able to bury their dead and achieve some measure of closure. With experiences like this, it did not seem far-fetched to believe that government might have deliberately flooded the 9th ward to protect people and property elsewhere, or that government had simply neglected to spend funds to maintain levees in the 9th Ward.

Among Katrina victims, as measured by three University of Chicago political scientists, the vast majority of African-Americans (71 percent) and a very small number of whites (3 percent) believe that blacks deserve reparations for slavery; support for reparations for Jim Crow is similar (74 percent in favor among African-Americans, 5 percent among whites) (Dawson *et al.* 2006). For African-Americans, reparations would provide equality, or at least a more equal starting point, for blacks to compete with whites in U.S. society. It does not take a long study of New Orleans to realize that African-Americans have not had an equal chance to succeed. Forty-four percent of African-Americans in the state are poor, as compared to 9 percent of whites. This raises a problem for the idea of social justice, or a Just City, as Susan Fainstein argues persuasively in favor of in this volume. Going against a powerful white majority disinterested in equality and opposed to equal opportunity to develop one's capacities (the actual goal of most black reparations demands) risks ostracism or censure of justice claims in state and national politics. Disregarding the claim for black reparations as "impractical," on the other hand, draws a cynical eye from African-Americans, who are well acquainted with equivocation about their justice claims. It may be that discourse and reason will not resolve this difference. While a Just City concept is appealing in its openness to various interpretations, even as it maintains a focus on disadvantage, and presents an opening for dialogue and a needed move away from rationalist-communication approaches, the racial difference will not be resolved without addressing structural problems that demands for political power, economic equality or reparations seek to address. For Katrina victims pressed for affordable housing and good schools in the wake of the storm, even softer terms like "equity" and "justice" met with a negative response. The substance of demands gives character to the words, not the other way around.

The University of Chicago survey also contained a surprise. While expressing a deep distrust for government, African-American adult victims of the disaster spoke very positively of their interaction with faith-based charitable organizations in the aftermath of the hurricane. Most black men and women remembered white volunteers in cities like Houston, Baton Rouge, and Tulsa as kind, responsive, and consistent. Federal bureaucracies, such as the Small Business Administration and FEMA, were characterized as cruel, uncaring, and incompetent. While black respondents remembered these volunteers with fondness, their interactions did not lead to shared perceptions of the political and racial implications of the disaster. This seems to suggest that blacks make a distinction between whites' individual attitudes (much progress) and the way politicians and other leaders manipulate race (very negative). This is an interesting point; it suggests a gap between southern white people with positive racial intentions and the governments they elect.

How could there be so many white Americans committed to improving race relations following Katrina, and yet so many immoral and malevolent government policies? In the 1960s, Martin Luther King, Jr., wrestled with a similar question. It is useful to recount this history. King regularly emphasized that millions of white people in the South wanted to end racial segregation, but they were timid in the face of governmental power. King believed that white citizens had to do more than perform individual acts of kindness toward African-Americans; they had to make government reflect their values —to become politicized citizens. "Government alone," he argued, "has the power to establish the legal undergirding that can ensure progress." Bad things happen in this country and in this world, he noted, when the good people of this country are silent about the evil they see around them and do nothing to convince their neighbors to act better—this was the essence of King's view on progressive politics. King also understood that society couldn't achieve peace, racial harmony, and genuinely integrated communities if some people are living with superfluous wealth while others live in abject poverty. During his lifetime, as today, many children living in the South did not eat three good meals a day. King emphasized that poor children do not deserve to suffer. Their suffering is unearned. King understood that society pays a terrible price for the resentment and despair brought about from unearned suffering. Reviewing King's message shows that rather than being an aberration, the unearned suffering of contemporary New Orleanians is part of a long history of ills in the U.S.—and also that there is a heavy price to pay for injustice.

When it came to politics, King knew that poor Southern whites were nearly as oppressed and under-educated as blacks. Historically, poor whites in the South had to compete for jobs with slaves or near-slaves, and this dragged down their income. Poor whites had always been told that they were better than blacks, because of their white skin. Many blamed blacks for major problems in society—if taxes were too high, blacks were the cause of it; if unemployed whites couldn't find jobs, it was because blacks were getting affirmative action; crime was essentially a problem of black behavior.

King believed that if blacks convinced poor whites that the two groups are stronger together than apart, and that racism cost them heavily too, eventually Southern whites would reject racist stereotypes and work to improve their lives together with blacks. This was the essence of Dr. King's dream. When Bill Rouselle said to black labor leaders that "this is the best opportunity we've had since Reconstruction," he meant that this is the best chance at such a poor white–black coalition (which would be an undefeatable coalition in Southern politics) in 150 years. There were efforts at constructing such a coalition, also including Latinos, in the wake of the storm.

LABOR MOVEMENT INTERVENTIONS

Leaders of U.S. labor unions, like many people worldwide, were outraged at what they saw on television following Katrina and sought to respond. A few national leaders had local affiliate unions in New Orleans and came under pressure from their members to help. Some labor leaders saw the crisis as an opportunity to build a new image for labor unions as organizations committed to the broader social good—building affordable housing, eliminating poverty, and fostering racial equality. They realized that it was especially significant to do this in the South, an anti-union "right to work" region. Historically in the South, unions were denounced for "race-mixing" until unions backed off aggressive organizing among African-Americans in the 1940s (Kelley 1994). Lacking a base among African-Americans—the poorest workers most interested in organizing—unions never generated enough of a movement to establish themselves in the region.

The AFL-CIO and, to a lesser extent, the Service Employees International Union, which led a split from the AFL-CIO soon before the storm, took an early and active role in trying to organize and rebuild New Orleans. The AFL-CIO's Investment Trust Corporation, a pension investment arm of the building trades, decided to invest $1.2 billion in rebuilding housing and starting new commercial ventures in the city. Their plan was to begin by rehabilitating public housing developments that suffered little damage during the storm. An immediate target was Lafitte Houses, a 900-unit public housing development near the French Quarter. Employers, such as the Avalon shipyards, were losing large contracts due to their inability to find workers (absent because workers lacked housing). The AFL-CIO wanted to use public housing temporarily to house workers. Once employed, most workers could move out of public housing and create a market for rehabilitating neighborhoods throughout the city. The AFL-CIO also planned to bring unemployed residents of the city into pre-apprenticeship ("job-ready") programs, apprenticeship programs, and eventually into unions. The AFL-CIO's pension funds would be targeted towards projects using union contractors, ensuring quality employment for newly-trained workers.

On June 14, 2006, the plan was outlined by John Sweeney, President of the AFL-CIO, at a press conference with Mayor Ray Nagin.[3] The mayor greeted it with great enthusiasm. Many blacks in the audience were stunned that white labor leaders would make a huge investment to bring back a mostly black city and target the black poor for economic advancement. Jerome Smalls, a veteran community activist in his late 70s, said that he had not seen anything like this in New Orleans since the Longshoreman organized interracially during his youth. He added that younger black people in New Orleans had absolutely no conception of a progressive inter-racial labor movement and suggested that the AFL-CIO do a publicity campaign on black radio and in black organizations to explain what they were doing (something the AFL-CIO never followed up on).

Problems soon emerged. Prospective workers entering job-training centers needed remedial education. A majority, as mentioned earlier, had 4th grade or less reading and math levels. Yet, carpenters and electricians require at least a 12th grade entry level for training. The AFL-CIO created the Gulf Coast Construction Careers Center, a three-week program that provides a three-week orientation and introduction to construction careers with corresponding classes in "mathematics, labor history, tool identification, and safety"[4] and life-skills training. But the AFL-CIO is not a substitute for public schools. It was (and is) apparent that the unions needed to join, or form, a broader political coalition to push for state and federal funding for remedial education; this has been an exceptional political challenge for the labor staff—many are outsiders to Louisiana politics. Unions are also confronting the widespread health problems and lack of access to transportation that has long characterized New Orleans' black community. Their graduates face stiff competition from non-union developers who rely on immigrant day laborers and cheap building materials.

Another problem was that the Bush Administration opposed labor's initial plan to renovate public housing. Soon after the AFL-CIO announced their intention to renovate public housing throughout the city, the federal department of Housing and Urban Development (HUD) announced its intention to tear public housing down in New Orleans. HUD's decision came despite the fact that several large public housing developments, such as Lafitte Houses, suffered minimal damage during the storm. The AFL-CIO's initial idea was to acquire site control of 200 vacant parcels in the neighborhood surrounding Lafitte Houses, to develop those sites as affordable housing, and then to offer former residents of Lafitte Houses the option of moving into newly developed sites in the surrounding neighborhood, known as Treme, or back into Lafitte Houses. The "right of return" would apply to the broader Treme neighborhood. Since surveys of HUD residents at the time suggested that only about half wanted to move back in Lafitte, it was anticipated that both Lafitte and the surrounding neighborhood would become more economically integrated than before the storm. With their homes slated for demolition, and no firm assurance of a right to return to the city, residents of public housing initiated multiple protests and filed suit in federal court. However, they lost the Court battle and HUD put strong pressure on city officials to support demolition. The buildings are being demolished.

Blacks and Latinos

The rebuilding of New Orleans' housing and infrastructure, if done with job-training and union-scale construction wages, could have led to a major improvement in living standards (and a likely reduction of crime) among the city's (mostly African-American) poor communities. Yet with limited federal assistance for reconstruction and a long anti-union tradition in construction, the door was wide open for the recruitment and abuse of undocumented immigrants (mainly Latino). Independent subcontractors recruited immigrants (estimates range from 30,000 to 100,000 workers), to do the bulk of the clean-up and reconstruction work. This posed a dilemma in a city populated by large numbers of unemployed African-Americans and aggravated tensions between blacks and Latinos on the streets (Browne-Dianis *et al.* 2006).

Hoping to develop inter-racial collaboration around issues of worker justice in the Gulf, Oxfam America (Oxfam) and the Unitarian Universalist Service Committee (UUSC) hosted a conference in Homer, Louisiana in April of 2006. The justice issues were of two types: young blacks being out of the workforce and Latino workers being abused on the job. African-American activists at the conference reminded participants that worker issues have been a key part of the ongoing racial struggle in the South. "Civil rights was about workers with no rights," said one speaker. Another noted that Martin Luther King, Jr., was murdered during a sanitation workers' strike in Memphis. Doris Koo, currently the President of Enterprise Community Partners, spoke of the race issue from a different perspective. She said that she was raised in a "dirt poor" community in Hong Kong, that her mother worked in a sweatshop, and that economic desperation and a lack of rights led her to come to the United States. She said that this was all part of the colonial experience under British Imperialism, and that current immigration from Latin America and Asia into the U.S. cannot be separated from U..S and European colonialism in those regions. She said that she was tired of the way that race was being discussed at the conference and in the country. Racial advocates in the U.S., Koo emphasized, must understand colonialism (and race in colonial history) in the same way that immigrants must understand slavery and the civil rights movement. Gerry Hudson, Executive Vice-President of the Service Employees International Union, and an African-American, added that immigrants and African-Americans share a history of exploitation at the hands of the same governments and corporate interests. If they do not understand each other's histories and struggles, then blacks, Latinos, and Asians will end up fighting each other over what amounts to scraps at the table of corporate wealth. The conference ended with these issues unresolved. Only three of the scores of attendees (mostly white) at the conference attended the workshop on "workforce development," focused on the issue of why blacks cannot gain jobs. This angered some of the African-American participants. As activist Charese Jordan put it: "It's hard to make workers rights a priority when we're not working" (Graham 2006).

The Homer conference was a snapshot of tensions rising between African-Americans and Latino immigrant workers in many parts of the country. This subject, as well as tensions emerging from middle class black and Latino competition, was a major topic of discussion during the Democratic Party presidential primaries, especially in Texas and California. In California, 80 percent of African-American voters supported Barack Obama, while nearly 70 percent of Latino voters supported Hillary Clinton. Clinton also enjoyed predominant Latino support in Texas. Media pundits have correctly portrayed relations between blacks and Latinos as uneasy and competitive. On the other hand, following in the tradition of earlier leaders such as Martin Luther King and Cesar Chavez, many activists in both communities maintain that a political disjuncture between blacks and Latinos could be fatal to the country's hope for a progressive movement. The reality is that both Latino and Black communities around the country are in peril. Latina women earn less than half of what white men earn. Only about half of young African-American men are working, and more than 10 percent are in prison. Added to this is the impact of rising energy costs, lack of health insurance, high rents, and mortgage foreclosures, and failing public schools.

Immigration muddies the water among the rapidly changing minority population in the United States. Employers purposefully recruit immigrants from Latin America, the Caribbean and Asia for low-wage, unsafe work in this country. In the 1940s and 1950s, employers were somewhat successful in undermining the then powerful labor movement by recruiting black workers out of the Jim Crow South. Rather than fight Jim Crow, organized labor mostly stood on the sidelines during the civil rights movement. This maintained conflicts between poor whites and African-Americans that the U.S. labor movement never overcame, and it is a leading cause of their weakened state in New Orleans and elsewhere in the South today.

It is now strategic for conservative forces to split African-American, Latino, and Asian-Pacific Islander voters. This trend will become stronger in coming years as more people of color enter the electorate. It is not enough to say that since minorities have similar interests—good jobs, quality education, and healthcare—they will unite. Poor whites and blacks have had similar interests for more than a century. Beyond the ignorance and stereotypes of the other, they have been distracted and divided by partial benefits such as white privilege and affirmative action, by the maneuvers of politicians dividing the electorate for advantage in elections, by employers, by middle-class aspirants of their own groups seeking jobs, grants, and contracts at the expense of the other group. All of these factors are at work in dividing African-Americans, various Latino groups, and various Pan-Asian groups. Uniting them is no simple matter; it requires leaders with a broad vision of racially and ethnically integrated communities. It requires supporting struggling workers in sweatshops across the Southern U.S. border, as well as building civic institutions to fight ruthless employers operating cross-nationally and opportunistic politicians "making hay" on the immigration issue. As opposed to a U.S., or Euro-U.S., concept of a Just City as Fainstein develops in her chapter in this volume, contesting demagoguery against immigrants requires that social justice planning address what Peter Marcuse in this volume calls "structural issues" that cross national borders, such as the existence of U.S. companies paying extremely low wages and dumping toxics (which municipalities do also) right across the Mexican border in cross-national cities such as Tijuana/San Diego. As mentioned earlier, such cross-national oppression is a prime structural cause of immigration. How to develop such leadership and vision among "minority" groups is as important a challenge as developing visionary political leadership in white communities.

PLANNING AFTER THE FLOOD

Disasters create opportunities to improve cities, and also opportunities for self-serving actions that do not help cities recover. In fact, these two impulses fight it out after disasters. The earthquake in Mexico City in 1985 gave rise to élite opportunism, followed by community resistance, and this put in place a democracy movement that eventually led to the overturn of the ruling party in Mexico that had governed the country since the beginning of the twentieth century. The riots in L.A. led to a lot of organizing and increased communications among and between community groups of different ethnicities in Los Angeles. It was a key factor in the transformation of L.A. politics that led to the election of its current Mexican-American Mayor, Antonio Villaraigosa (a labor and community activist) and to Karen Bass (a militant African-American community leader) as Speaker of the California Assembly.

There were two opposite movements immediately following the flood in New Orleans. One was (and is) an attempt by city élites to keep poor people from returning to the city by building golf courses on the terrain of poor neighborhoods. Another was a movement of community based organizations and grassroots activists. Dozens of local organizations convened survivors of the storm and plotted strategies to restore housing, schools, health services, to clean up piles of debris, and to contact other survivors scattered throughout the country. Several foundations supported these efforts by funding community design and planning processes encouraging residents to envision how their neighborhood would recover. Most of the community organizations were very small, unincorporated as non-profit organizations, disconnected from their scattered pre-storm members, and their members in the city were usually struggling with their own personal losses from the storm. Only a few of the organizations had the capacity to meet regularly with foundation-hired planners or to work systematically on neighborhood rebuilding. Those that did created many new innovative building designs and appealing streetscapes. There is a sense that some devastated neighborhoods, such as Broadmoor and parts of the 9th Ward, are slowly coming back to life. This is valuable work, not to be discounted.

On the other hand, without substantial federal support to rebuild affordable housing for the tens of thousands of displaced residents, most neighborhoods are not coming back. If they do, the mostly black people who lived there before, the people that gave New Orleans its unique character will not populate them. Indeed, much of what is called "New Orleans style" derives from traces of African and Caribbean heritages that were better preserved in New Orleans than elsewhere in the country. This was a strong cultural bond that held residents in the city together. Prior to the storm, close to 90 percent of black New Orleanians were born there, the highest native-born black population of any large city in the U.S. Failing reversal of federal policy, that New Orleans will be gone for ever.

More than anything, community organizations following the storm needed a unified plan, and ways and resources to collaborate in pressing for state and federal support. While there was substantial outside private and philanthropic support for rebuilding efforts in New Orleans, few emphasized building a political movement in the city. Instead, foundations encouraged competition among neighborhood activists for funds to support neighborhood revival. While this promoted progress in isolated pockets of the city, there were winners and losers in these competitions, the overall result has been a perpetuation of political fractures—something that has plagued community groups in New Orleans for decades.

Urban planning usually fell into two categories. On one side were grandiose plans for rebuilding a new transit system or other infrastructure that called for massive political support for rebuilding the city, yet did so at a time when there was weak federal or state support for even emergency aid. Planners did not address the imperative of building political support for making such public investment viable. Other planners, rather than establishing a citywide agenda and unified political movement for securing adequate resources to rebuild the city, encouraged residents to think small, in terms of their own immediate neighborhood. As noted earlier, resident groups became competitors for extremely scarce resources. Another striking issue was that planning was fixated on housing. While housing was clearly an immediate need, planning processes seldom addressed the core problem in sustaining most low-income neighborhoods, or in enabling dispersed residents to return, which is quality work. Even if there were increased federal and state aid to rebuild New Orleans, there is no system in place to ensure that displaced unemployed residents would be trained or hired, or that these jobs would pay union-scale wages (i.e. living wages). In the absence of any significant plan for job training and economic improvement at the bottom-end of the labor market, many community leaders and city officials publicly sought to block poor people from returning to the city. While part of their resistance may be rooted in prejudice against poor people, another part was undoubtedly based on observation and fear of drug-dealing and violent crime—things closely associated to long-term unemployment. All told, so-called professional planners tended to suck the politics out of an explosive political situation and marginalized themselves in the process. While many planning students and community activists tried to create citywide political coordination with few resources aside from their personal commitment to social justice in the city, most planning initiatives distracted residents from urgent political imperatives.

NEW ORLEANS AND THE NATION

New Orleans presents a paradox. On the one side, the experience since the storm shows promise of a new America. Here in a deep southern region, legions of ordinary white citizens reached out to support thousands of desperate African-Americans. Poor white survivors of the storm realized, in many instances, that they are no more secure in the U.S. than are blacks, and there have been more multi-racial organizing efforts since the storm than in decades. Hundreds of dedicated white activists—planners, environmentalists, community organizers—have moved to the city to play a part in rebuilding it in a way that allows former residents to come home. The youthful interracial activism is reminiscent of the civil rights movement of the 1960s, and again provides hope that racial divisions can be overcome. Labor unions, unlike the 1960s, are putting resources into New Orleans in an attempt to play a part in rebuilding the city and an inter-racial labor movement.

Yet, the overall outcome has been negative, even tragic. New Orleans needed a movement. It was apparent that after decades of building expertise in housing development, planners did not know how to organize politically. After decades of fighting each other for scarce government and foundation resources, community organizations were more skilled at mutual destruction than collaboration. There are more Latinos in New Orleans than before, yet Latinos and blacks do not yet know what to make of each other. Construction unions moved tentatively beyond their narrow focus on middle-class jobs for white suburban workers, but were unprepared for movement building. Local government, isolated in state and federal political circles, was broke and ineffective.

As the nation and the world moves closer to environmental catastrophe, New Orleans stands as a frontline example of what may happen elsewhere when city infrastructures prove incapable of handling the effects of weather or ecological conditions. Low-income populations are the most vulnerable, and élite interests will likely act opportunistically. There will be unprecedented opportunities for building powerful cross-racial movements, yet without attention to popular education to establish a social and historical context for present dilemmas, or leadership development beyond technique, or new models for resource development and sharing, that potential will not be realized.

NOTES

1 Thanks to Justin Steil for his patience and insightful comments.
2 Author's conversation with Mayor Nagin and aide, Winter 2006.
3 The AFL-CIO press release is available at: www.aflcio.org/mediacenter/prsptm/pr06142006.cfm
4 See: http://neworleanslabormedia.org/katrina-survivor-pursues-boilermaker-career-0

REFERENCES

Barry, J. (1997) *Rising Tide: The Great Mississippi Flood of 1927 and How it Changed America*, New York: Simon & Schuster.

Browne-Dianis, J., Lai, J., Hincapie, and Saket Soni, M. (2006) *And Injustice for All: Workers' Lives in the Reconstruction of New Orleans*, Washington, D.C.: The Advancement Project.

Dawson, M., Lacewell, M., and Cohen, C. (2006) *2005 Racial Attitudes and the Katrina Disaster Study*, Chicago: University of Chicago.

Dyson, M. (2006) *Between Hell and High Water*, New York: Basic Civitas.

Fairclough, A. (1995) *Race & Democracy: The Civil Rights Struggle in Louisiana, 1915–1972*, Athens: University of Georgia Press.

Foster, R. and Giegengack, K. (2006) "Physical constraints on reconstructing New Orleans," in Wachter, E. and Wachter, S. (eds) *Rebuilding Urban Places After Disaster*, Philadelphia: University of Pennsylvania Press.

Graham, L. (2006) *Convening on Workers' Rights in the Gulf Coast: Understanding and Connecting Organizational Activity since Hurricane Katrina*, Hammond, Louisiana: Unitarian Universalist Service Committee.

—— (2008) *The Right to Work for Less: The AFL-CIO, Unionism, and Green Collar Jobs in Post-Katrina New Orleans*, Cambridge, MA: MIT.

Johnson, C. (2005) "Police made their storm misery worse," *San Francisco Chronicle*, September 9; B1.

Kelley, R. (1994) *Race Rebels: Culture, Politics, and the Black Working Class*, New York: Free Press.

Lerman, H., Holzer, J., and Robert, I. (2006) *Employment Issues and Challenges in Post-Katrina New Orleans*, Washington, D.C.: The Urban Institute.

MSNBC (2005) "New Orleans murder rate on the rise again," August 18, available at: http://www.msnbc.msn.com/id/8999837 (accessed August 20, 2008).

Rouselle, B. (2006) Personal conversation between Bill Rouselle and the author, June.

SATYA (2005) Going home: the Satya interview with Dr. Beverley Wright, available at: http://www.satyamag.com/nov05/wright.html (accessed on August 23, 2008).

Conclusion

Just City on the horizon: summing up, moving forward[1]

Cuz Potter and Johannes Novy

> The idea of justice is, like other moral ideas, not imparted to men by some revelation, and just as little is it an arbitrary invention; it is the necessary product of our moral intuition and our logical thinking, and in so far it is an eternal truth, manifesting itself [in] ever new yet ever similar metamorphoses.
>
> Gustav Schmoller (1894: 40)

> A genuinely humanising city has yet to be brought into being . . . to an urbanism appropriate for the human species. And it remains for revolutionary practice to accomplish such a transformation.
>
> David Harvey (1973: 314)

The paradox of our time is that the unprecedented capacity to secure human well-being should generate such precarious existences for the vast majority of humanity (Brodie 2007: 104). What is to be done to change this situation? How can urban and regional planning contribute to a more socially Just City and how might social theory assist? Questions like these prompted Susan Fainstein to formulate her notion of the Just City, which inspired this volume. Fainstein has built on the foundational works of scholars like Harvey (1973) and Lefebvre (1968/1996) and on the work of numerous other researchers pursuing similar issues. While many of these thinkers, seeking a more appealing future, focused on a radical critique of the existing urban order and the conditions responsible for it, Fainstein's writings on the Just City attempt to move beyond critique.

"Many scholars in the political-economy tradition offered critique without formulating specific criteria of what was desirable," Fainstein (2001: 885) argued in one of her earlier texts on the Just City. But utopian urbanism does not satisfy Fainstein either: "While utopian ideals provide goals toward which to aspire and inspiration by which to mobilize a constituency, they do not offer a strategy for transition within given historical circumstances" (Fainstein this volume: 28). Instead, she argues that progressive urban change requires transition strategies that take account of both the desirable and the feasible.

The chapter that opens this volume represents Fainstein's most fully elaborated effort to date to develop criteria for a just city and to again position urban social justice as a central goal of urban planning. The contributions to this volume can be loosely categorized as developments of and challenges to her formulation. In the following sections, we identify contributors' practical applications of the Just City concept, and issues that have not yet been sufficiently addressed by the concept, or theoretical challenges to the just city. We conclude by questioning the just city's reliance on working within existing structures towards change as a necessary prerequisite to garner the support of the middle class.

SUMMING UP: DEVELOPMENTS

Fainstein's conception of the just city accepts the global capitalist political economy as a theoretical frame for policy, while opposing the more egregious outcomes of the dominant, neoliberal frame of policy making. Though Fainstein's project is ultimately directed toward defining the just city, at its current stage of development she does not seek to define the just city so much as she seeks to explore the implications of well-established theories of justice for formulating urban policy. Though these theories themselves may contradict each other (cf. Nussbaum 2006), Fainstein problematizes aspects that "are also of central importance to urbanists and can therefore be extended to evaluating urban policy" (p. 25). Authors develop the implications of three sets of common planning issues—collective rights and goods, process versus outcome, and the tensions between growth, equity and diversity—and question her reliance on Amsterdam as a model for the just city.

Collective rights and goods

Fainstein argues that in addition to the individual liberties and individual primary goods that form the basis of contractarian theories of justice, the just city must also consider collective rights and collective goods in formulating policy. While the focus of contractarian theories of justice are generally based on leveraging individual (property) rights for individual gain, Fainstein notes that this perspective ignores nonmaterial causes of social inequity, like racism and sexism, and desirable collective goods like social well-being. Instead, she insists that any régime of justice must account for them by acknowledging "the coherence of collectivities and their structural relationships to each other" (p. 29). "[T]o achieve a better city in a better society, we need something more than justice for particular parties," Marcuse (p. 101) concurs, "We need to deal with the ownership, control and use of the commons." Harvey (p. 48) also argues that the right to the city must go beyond individualized rights to shape a collective politics oriented around social solidarities, identifying a "vast range of innovations and experiments with collective forms of democratic governance and communal decision making." Connolly and Steil's study in this volume, which elaborates one such effort, suggests that representatives of sociospatial collectivities like "communities" and "neighborhoods" can institute their collective rights in decision making through legislation. And DeFilippis highlights the role of immigrant worker centers, worker cooperative businesses, and strikes and campaigns. These two pieces reflect Fainstein's conclusion that: "[I]n the new century, effectiveness probably means organizing around work status when it overlaps with racial, immigrant or gender situation" (p. 34). While this may be the case in developed countries, Maricato's record of the long fight for the right to the city in Brazil illustrates that much of the developing world's collective struggle still centers around economic class and can ultimately flower into the legislation of new planning laws that alter the nature of property rights and distributional justice, such as the "right to the city" legislation passed in Brazil in 2001. Two caveats are offered to this generally optimistic prognosis, however. First, Maricato emphasizes that even incremental collective gains are under threat from conservative, capitalist forces, requiring ongoing vigilance and struggle. Second, in their discussion of the production and reproduction of hierarchies of recognition, spaces, and ethnicities in Israel, Yiftachel, Goldhaber, and Nuriel point out that recognition of collectivities does not necessarily produce justice. Within different political contexts and projects, they argue, affirmative recognition or egalitarian indifference can disguise more cynical political motives that generate discriminatory policies.

Process and outcome

For Fainstein, it is vital to attend to both process and outcome, as just processes do not guarantee just outcomes and vice versa. Though, as Marcuse (p. 96) points out, more work remains to be done to further connect process and outcome, Fainstein's emphasis on the role of counter-institutions "capable of reframing issues in broad terms and of mobilizing organizational and financial resources to fight for their aims" (p. 35) in generating social change resounds throughout the volume, as the examples in the preceding paragraph demonstrate. She also suggests that a "broad-based media [must] exist to communicate alternative approaches" to support such counter-institutions (p. 34). Notably, Wolf-Powers' contribution explores the relation between media and counter-institutions through the example of local cable programming and a planning newsletter in Bedford-Stuyvesant, Brooklyn, in the 1960s and 1970s. She argues that state support and planner interventions can improve process by creating and maintaining counter-publics, or marginal publics, in order to frame issues in ways that not only critique existing policies and plans but also assist in the development of alternatives. Connolly and Steil also emphasize the inseparability of process and outcome, providing an example of how less powerful organizations can shift the contours of discourse and power to favor local voices in decision making by employing legislative changes and forming alliances with intermediary organizations that "speak everyone's language." Moving the debate to a higher level of abstraction by discussing the contentious relationship of "planning theory focused on communicative interactions and those [scholars] pursuing a more critical political-economic approach," Fischer in his contribution posits that both the "more traditional political-economic orientation" with its focus on planning outcomes and the discursively oriented communicative approach are "at important points necessarily dependent on one another."

Growth, equity, and diversity

Several contributors identify potential directions for addressing the tensions between growth, equity and diversity. Refusing to reduce the "urban question" to either political economy or culture, Fainstein discusses Young's idea of "differentiated solidarity" as a "realistic approach to the issue of multiculturalism" and stresses the potential of Young's work as a guiding framework for Just City adherents when striving to move towards forms of urban development that embody heterogeneity, cultural mix, and fusion as well as equity and fairness.

Thompson's discussion of cooperation and conflict between union labor, undocumented workers, and local communities in New Orleans' effort to recover from Hurricane Katrina illuminates the need for greater levels of differentiated solidarity and recognition among these urban groups more generally if shared and individual goals are to be achieved. DeFilippis concludes from his study of labor in New York City that growth and equity are not mutually exclusive concepts and that demands must be made directly on capital rather than the state. He argues that industry-specific worker organizing campaigns and institutions like workers' centers that combine workplace issues with community and cultural issues are crucial to giving workers greater control over labor conditions. Finally, Connolly and Steil's discussion of brownfield redevelopment in the South Bronx suggests that community organization around common concerns can overcome ideological differences and environmental racism while maintaining sustainable development through collective political representation. In sum, the pieces in this volume point to the situated nature of balancing equity, diversity, and growth. Whereas some suggest that localized interventions are critical and possible, others, highlighting particular conflicts between the goals of growth and equity, emphasize trade-offs or clashes among them.

Concrete model

Fainstein also revisits her consistent advocacy for Amsterdam as "an actual model of social justice" (p. 32) to guide other localities in developing just policies and governance. In noting the city's rising levels of intolerance and fear of difference as well as other regressive trends, she is beginning to move away from this position (this volume). Novy and Mayer make two arguments for moving even further away. They first suggest that identifying a particular city aggrandizes circumstances in that city and dulls the local political impetus to push for important changes. Second, they argue that their examination of the growing impacts of creeping neoliberalism in the Netherlands demonstrates that the model no longer represents an ideal. Skeptical of the just city formulation's "current focus on (and problematic dualism between) North American and European cities," they suggest that rather than looking for inspiration in Amsterdam (and perhaps other European cities) where conditions are declining, "the search for models of transformative urban action should be extended into other parts of the world to inform and broaden the scope for progressive activism in American and other 'first world' cities" (p. 106). From his postempiricist position, Fischer (p. 60) likewise argues that "one can offer a theory of the Just City, but it cannot be more than one of numerous other contested positions and will be treated as such by those with different preferences. This is to say, it cannot be established once and for all by accepted criteria."

SUMMING UP: FUTURE DEVELOPMENTS

While the previous section identifies a number of ways concerned urbanists can work toward the Just City as it is currently framed, the narrow range of contributions we were able to include also highlight directions in which the search for the Just City can expand to achieve its potential. The following section addresses some of the voids in the current Just City formulation and points to issues that need further development: scale, spatiality, environmental sustainability, intertemporality, and gender and sexuality.

Scale

In her emphasis on the capacity of city government to implement change, Fainstein's representation of the tension between growth, equity, and diversity masks the deeper issue of the appropriate scale for challenging contemporary processes of urban and regional development. Though DeFilippis identifies significant opportunities at the urban scale to establish living wages for workers in nontradable (generally service) industries, Fainstein herself states, "justice is not achievable at the urban level without support from other levels" (p. 21). In response, Marcuse (pp. 100–101) argues that city government is too limited to create a Just City and that solutions and causes must be sought at larger geographical scales. Addressing the disaster in New Orleans, Thompson, for example, observes that establishing just relations among the ethnic and community groups in a single urban setting requires an understanding of the long, transnational history of colonialism and global racism.

Maricato also illustrates these broader causal relations by elaborating the connection between contemporary neoliberal "financial colonialism" administered by core, developed countries and unjust conditions in peripheral countries of the developing world. The inevitable conclusion is that just situations at one scale may be unjust at others, as Massey (2005) has pointed out. Critiquing the Russian doll geography of ethics, care, and responsibility (2004, 2005: 186), in which local population, environment, economy, etc. are prioritized over larger scale loyalties and responsibilities, Massey argues for reframing spatial politics in relational terms.

For Massey, spatial relationality demands that we rethink our geographical responsibility for distant places. In light of the increasing interconnectivity and interdependencies of spaces and places under globalization, future discussions of the Just City should elaborate a perspective on space and spatial scales as socially produced and reproduced that is capable of addressing not just local but also international, inter-regional, and intra-regional justice relations. This framework should synthesize the local as the cradle of all concrete processes and interventions with a "global sense of place" (Massey 1994), which includes extra-local and non-territorial conceptualizations of social justice as well as a recognition of our responsibilities to others who are connected to aspects of our identity, including those related to our places. A more comprehensive vision of the Just City would combine a concern for the just treatment of an urban area's inhabitants and visitors with a consideration of the extraterritorial responsibilities of that locality, instead of drawing boundaries that artificially wall them off from the wider world. (See Fraser 2005 for a preliminary effort to define justice relationally.)

Spatiality

There is an even deeper spatial concern, however. Due to its primary focus on distributive and procedural justice, the Just City has relatively little to say about space's role in producing injustices nor about the physical form of just spatial development. Pirie (1983: 471) was among the first to ask if space can be just or unjust and concluded that if space is conceived as a container of social process, "spatial justice" is simply shorthand for "social justice in space." Building on the work of Pirie, Harvey, Lefebvre, and others, Dikeç (2001: 1792; see also this volume) argues for treating spatiality and justice as a dialectical relationship, developing a focus on both "*the spatiality of injustice*—from physical or locational aspects to more abstract spaces of social and economic relationships that sustain the production of injustice —and the *injustice of spatiality*—the elimination of the possibilities for the formation of political responses." He argues that taking the production and reproduction of spatial "permanences" into account would assist in exploring the dynamic processes of social, spatial, economic, and political formations and in determining their justice or injustice. In short, Dikeç's approach would enrich the Just City concept by spatially situating social processes, including planning techniques, in a way that reveals how prior spatial productions may perpetuate or be used to perpetuate just or unjust outcomes.

Environmental sustainability

Closely related to the issue of geographical scale, a further issue this volume only peripherally touches upon, is environmental degradation (Agyeman and Evans 2003; Agyeman 2005). Environmental concerns, as well as issues of justice and equity are linked and interdependent locally, nationally and globally at the levels of both problem and solution, yet exist in a complex and often contentious relationship. As Marcuse notes: "Programs and policies can be [environmentally] sustainable and socially just, but . . . they can also be [environmentally] sustainable and unjust" (Marcuse 1998: 103). While environmental considerations are raised in both Fainstein's initial work on the Just City as well as other chapters in this volume (see particularly Connolly and Steil), their role within the Just City concept remains underdeveloped. Connolly and Steil's piece on brownfield redevelopment in the South Bronx begins to explore the importance of the environmental justice perspective to the just city debate and suggests the need to take a fresh look at the synergies —and conflicts—between the goals of sustainability and justice. Meanwhile, little or no theorizing has been done on broader geographical problems like climate change and transboundary pollution that often originate in and descend upon urban environments, though the relational approach advocated by Massey may provide a point of entry to analyzing these issues. With the world population growing exponentially and environmental deterioration accelerating, environmental health and justice are becoming ever more pressing, and there is no doubt that future developments of the Just City will be incomplete if they do not make the reconciliation of issues related to social justice, environmental protection, and sustainability both a practical and theoretical focal point.

Intertemporality

Most prominently defined by the Brundtland Commission (WCED 1987) as "meeting the needs of the present without compromising the ability of future generations to meet their own needs," the notion of sustainability challenges the current Just City formulation beyond strictly environmental concerns by raising the issue of inter- and intra-generational justice (Laslett and Fishkin 1992). To what extent must we account for the effects of our actions on future generations and should members of one generation compensate others for past injustices? What are the obligations of old to young and young to old? Within planning, an immediate point of investigation would be how urban planning and policy can justly accommodate the needs of cities' aging populations. Given the world's ongoing demographic shift, particularly the aging of populations in Western countries and China, this issue should be directly addressed in future discussion of diversity and recognition in the Just City.

Gender and sexuality

Two other markers of difference that call for fuller development are gender and sexuality. Feminist research on planning issues, such as home, work, childcare, transportation and other forms of access, social services and more recently the moral construction of public space, extends the debate about justice and cities with tools for interrogating and critiquing the spatial expression and perpetuation of male domination and female subordination (Dias and Blecha 2007). Meanwhile, the literature growing out of geographical studies of gay and lesbian experiences and studies of sexuality and space more broadly challenges the "heteronormativity of space and the many ways in which everyday spaces reinforce the invisibility, marginalization, and social oppression of queer folk" (Brown *et al.* 2007: 8). Dikeç (this volume) pushes these notions further, arguing for Lefebvre's *right to difference*, in which justice arises not from the particularities that form the core of identity politics or stereotyping upon which recognition claims are made, but rather from escaping the trap of established categories of thought through *differing* in action. This approach contrasts with Young's (1990) work, which embraces identity and which Fainstein has attempted to incorporate by situating respect for diversity at the heart of the Just City concept. Thus, the inquiry her work has provoked might benefit from an expansion of its empirical and theoretical terrain by incorporating more of the unique contributions of feminist and queer perspectives, including the key insight that much that is written about space and its construction, often unconsciously, continues to be guided by a patriarchal and heteronormative ontology that requires critical deconstruction.

MOVING FORWARD: REALIZING THE JUST CITY

Following Fischer (this volume), the Just City formulation is maybe best understood as "an invitation to engage in a discourse—that is, to deliberate about the nature of equality and opportunity in a particular society and how its members might go about changing existing arrangements." As Fainstein posits in her contribution (p. 35), "by continuing to converse about justice, we can make it central to the activity of planning . . . , change popular discourse and enlarge the boundaries of action." For Fainstein, changing discourse involves developing alternatives to conservatives' and neoliberals' own vision of social justice, which posits that a just society can best be advanced by maximizing entrepreneurial freedoms within an institutional framework of private property rights, individual liberty, free markets, and free trade (see Harvey 2006; Brodie 2007). It also involves engaging more intensively with the middle class and their preferences. Fainstein (2000: 469) argues that a "persuasive vision of the Just City," if it is to mobilize large numbers of people, must not ignore "the interests and desires of the vast middle mass" and its evident trust in existing social relations. Visions that do otherwise, i.e., those that challenge the basic tenets of free market capitalism and the real or perceived benefits it entails, would alienate broad swaths of the public. It is primarily for this reason, it seems, that Fainstein is adamant that we must limit ourselves to change *within* the capitalist world-economy, i.e., to addressing what is currently possible within the present socioeconomic organization of society to achieve more just processes of urban development. Contrary to this perspective, several contributors to this volume argue that the progressive change for which the Just City formulation aims simply cannot be achieved without transforming the underlying economic structure. Novy and Mayer are "skeptical of [Fainstein's] assumption that urban social justice and a capitalist order can go hand in hand" (p. 116). And Harvey argues that: "The bundle of rights and freedoms now available to us, and the social processes in which they are embedded, need to be challenged at all levels. They produce cities marred by inequality, alienation, and injustice" (p. 45). In this view, not imagining alternatives predestines us to failure.

Does imagining alternatives to the existing economic order jeopardize the public support needed for progressive, democratic change, and must the Just City project therefore remain locked within the straitjacket imposed on conventional mass politics to have practical effect? We are not sure. First, though it may seem quixotic to attend to contestation when neoliberalism appears dominant, it could also be argued that it is particularly important at such times for progressives to imagine and create alternative (urban) futures. Second, from the early 1990s through the preparation of Fainstein and others' contributions to this book, global free-market capitalism seemed poised for a victory lap and opposition scarce. As Thomas Friedman (2000: 104) wrote, "Today there is no more mint chocolate chip, there is no more strawberry swirl and there is no more lemon-lime. Today there is only free-market vanilla and North Korea." Not anymore. Particularly in light of the current global economic crisis, it seems that more and more people are realizing that the neoliberal project has failed them by failing to deliver on its core promises of wealth, wellbeing, and freedom. They are joining the ranks of a variety of political and social movements that have long challenged the legitimacy and stability of the *status quo* by developing alternative imaginaries and practices that oppose the manifold injustices that dominate our lives. Such contestations reveal that many of the claims that have grounded the current global economic order are not so much a description of how the world is, as an image in which the world is being made (Massey 1999: 40). Cracks, as Leyshon and Lee (2003: 3) put it, have begun to appear in the edifice of the present neoliberal political economy, cracks that are opening new opportunities for propelling cities—and society at large—toward a better future for all.

While Fainstein (2000: 467) has argued that "creating a force for change requires selling a concept," in her most current proposals she effectively treats preferences in the market of ideas as fixed. Assuming that the middle and working classes' support of the existing economic order is given, she has developed the Just City to appeal to this fixed set of presumed preferences. However, she also notes in this volume that preferences are shaped by the widely promulgated set of beliefs and values society takes for granted, and she has previously employed Mannheim's work to posit a "rational subject who is capable of making comparisons and learning" within a given historical situation, i.e., a subject capable of acquiring perspectives broader than his or her own particular social standpoint (Fainstein 1996: 37). While Fainstein dismisses the Marxist thesis that material and institutional processes in capitalist society create false consciousness and urges Just City adherents to take the—real or perceived—benefits of a capitalist world economy seriously, she nonetheless also emphasizes the power of discourse to shape perceptions and preferences.

Given the pressing demand for alternatives to a system that has suffered catastrophic collapse, we believe that future debates surrounding the Just City formulation should explore the potential of utopian imagination in general and transformative utopian urbanism in particular. This is not to propose fixed solutions, as we have seen too many such "utopias of spatial form" fail in the past (Harvey this volume: 46; Harvey 2000: 160). Rather, utopian conceptions could be employed to overcome self-limiting resignation to current unjust social arrangements, to better pinpoint how and why these arrangements fail us, and to guide us towards our hearts' desire: the Just City (cf. Pinder 2005). Utopian thought can function "as both a diagnostic instrument of the constraints of the present and as a vision of an unrealized—and perhaps unrealizable—future" (Cevasco 2007: 125–126). This, as Eduardo Galeano reminds us, arguably represents its greatest political potential: "Utopia is on the horizon: when I walk two steps, it takes two steps back . . . I walk ten steps, and it is ten steps further away. What is utopia for? It is for this, for walking" (Galeano and Borges 1997).

NOTE

1 Special thanks to David Madden for his timely, useful, and inspirational comments.

REFERENCES

Agyeman, J. (2005) *Sustainable Communities and the Challenge of Environmental Justice*, NYU Press, New York.

Agyeman, J.B.R. and Evans, B. (eds) (2003) *Just Sustainabilities: Development in an Unequal World*, MIT Press, Cambridge.

Brodie, J. (2007) "Reforming Social Justice in Neoliberal Times," *Studies in Social Justice* 1(2): 93–107.

Brown, G., Browne, K., and Lim, J. (2007) "Introduction, or Why Have a Book on Geographies of Sexualities?," in K. Browne, J. Lim, and G. Brown (eds) *Geographies of Sexualities: theory, practices and politics*, Ashgate, Burlington.

Cevasco, M.E.B.P.S. (2007) "Archaeologies of the Future: Western Marxism Revisits Utopia," *Situations*, II: 120–127.
Dias, K. and Blecha, J. (2007) "Feminism and Social Theory in Geography: An Introduction," *The Professional Geographer* 59(1): 1–9.
Dikeç, M. (2001) "Justice and the spatial imagination," *Environment and Planning* A 33(10): 1785–1180.
Fainstein, S. (1996) "Justice, Politics, and the Creation of Urban Space," in A. Merrifield and E. Swyngedouw (eds.) *The Urbanization of Injustice*, Lawrence and Wishart, London.
—— (2000) "New Directions in Planning Theory," *Urban Affairs Review* 35(4), 451–478.
—— (2001) "Competitiveness, Cohesion, and Governance: Their Implications for Social Justice," *International Journal of Urban and Regional Research* 25(4): 884–888.
Fischer, F. (2003) *Reframing Public Policy: Discursive Politics and Deliberative Practices*, Oxford University Press, London.
Fraser, N. (2005) "Reframing Justice in a Globalizing World," *New Left Review* 36(1), 1–19.
Friedman, T.L. (2000) The Lexus and the Olive Tree: Understanding Globalization, Anchor Books, New York.
Galeano, E. and Borges, J.F. (1997) *Walking Words*, W.W. Norton and Company.
Harvey, D. (1973) *Social Justice and the City*, London: Edward Arnold.
—— (2000) *Spaces of Hope*, University of California Press, Berkeley.
—— (2006) "Neo-liberalism as creative destruction," *Geografiska Annaler* 88 B (2): 145–158.
Laslett, P. and Fishkin, J. (1992) *Justice Between Age Groups and Generations*, New Haven, CT: Yale University Press.
Lefebvre, H. (1968) *Le droit à la ville*, Paris: Anthopos, in H. Lefevbre (1996) *Writings on Cities*, trans. E. Kofman and E. Lebas, Oxford: Blackwell.
—— (1991) *The Production of Space*, Blackwell, Cambridge.
—— (2003) *The Urban Revolution*, University of Minnesota Press, Minneapolis.
Leyshon, A. and Lee, R. (2003) "Introduction" in A. Leyshon, R. Lee and C.C. Williams (eds.) *Alternative Economic Spaces*, Sage: London.
Mannheim, K. (1936) *Ideology and Utopia*, Harvest Books, New York.
Marcuse, P. (1998) "Sustainability Is Not Enough," *Environment and Urbanization* 10(2), 103–112.
Massey, D. (1994) *Space, Place and Gender*, University of Minnesota Press, Minneapolis.
—— (2004) "Geographies of Responsibility," *Geografiska Annaler B* 86B(1): 5–18.
—— (2005) *For Space*, London: Sage.
Pinder, D. (2005) *Visions of the City: Utopianism, Power, and Politics in Twentieth-century Urbanism*, Routledge, New York.
Pirie, G. (1983) "On Spatial Justice," *Environment and Planning A* 15(4), 465–473.
Schmoller, G. (1894) "The Idea of Justice in Political Economy," *Annals of the American Academy of Political and Social Science* 4: 1–41.
Soja, E.W. (1989) *Postmodern Geographies: the Reassertion of Space in Critical Social Theory*, Verso, New York.
United Nations World Commission on Environmental and Development (WCED) (1987) *Our Common Future*, Oxford: Oxford University Press.
Young, I.M. (1990) "City Life and Difference," in Scott Campbell and Susan S. Fainstein (eds) (2003) *Readings in Planning Theory*, Blackwell, Oxford.

Postscript
Beyond the Just City to the Right to the City

Peter Marcuse

Just City ideas are creatures of their times. Times, of course, are changing. The 1960s, and in particular 1968, symbolize a major change. This era opened a range of demands that include but go beyond justice. In theory, these demands are incorporated in the Right to the City. Despite a post-1968 conservative retreat of planning, the Right to the City has been developed in practice with concrete demands, some of which rely on traditional notions of justice; others point beyond, towards the new possibilities opened by the 1960s. The next task might be working out theoretically the bases for the new demands of the Right to the City movements. There are strategic political reasons to do so today, joining the interests of the Excluded with those of the Included.

THE NEW HISTORICAL SITUATION

Major changes affecting the urban landscape have taken place in the last 50 years, and they impact any conception of what a just or a good city should be like. These include changes in the political balance between competing forces, in the technologies of production, and in the role of governments. Such changes need to be a part of any discussion of the Just City. Goals for human activity are, after all, not manufactured at will by intellectual effort, but are created out of the challenges that are presented in real life, out of the problems and possibilities of the age. As David Harvey has written: "Concepts of social justice and morality relate to and stem from human practice rather than [from] arguments about the eternal truths to be attached to these concepts" (1973: 15).

While the search for an ideal city, for an ideal organization of life and of space, seems to be timeless and universal, analogous to the search for eternal truths, even eternal truths change in the meanings assigned them by men and women at different times and places. The suggestion in this concluding section is that there has indeed been a change in the meaning of the Just City, as a representation of that eternal search today, and that change has come about because of an historic alteration in what David Harvey calls "human practice."

The change can be dated and described in very concrete historical, economic, political, and sociological terms. Its origins lie in the industrial revolution and the rise of capitalism, and it has proceeded at an accelerated pace into the twentieth century. It has led to the widely acknowledged shift from its earlier stages to what is today variously called globalization, post-Fordism, a network society, a post-industrial economy, even the end of history, at least of a particular stage of history. Just as important for our purposes here, this change is reflected in altered political and cultural forms and actions, and specifically in the world-wide protest movements of the 1960s, the anti-colonialism struggles, the civil rights movement in the United States that was tested in universities and on the streets world-wide, the ideological debates around the New Left, and the divided working class, partially in support of change, partially encumbered by old practices and bureaucracies.

The connection between economic and technological changes and the cultural parallels that come alongside them lies in the possibilities opened by advancing technologies. Those advances have on the one hand opened a vision of a society of plenty in which a rational approach could be expected to satisfy all human needs with a minimum of mindless labor, in which poverty and want can be abolished and individuals can be free to develop their own potential securely and without worrying about survival. On the other hand, those very advances have increased the possibilities of social manipulation and social control, of manufacturing needs and distorting values, of maintaining allegiance to an existing system driven by the aggressive competitive pursuit of wealth and power.[1]

The conflict between these two aspects of the changed society has produced, although not always explicitly, a change in the goals of social action, in the visualization of the ideal towards which social action strives, a change that must lead us "beyond the Just City." The change requires us to embrace concepts such as an end to the acceptance of violence, racial, ethnic, and gender discrimination, consumerism, the competitive pursuit of wealth and power, false virility, hypocritical sexual mores, environmental degradation, commercialization of art and imagination; in short, an end to the production of one-dimensional people (a phrase my father coined in 1964; see Marcuse 1972). It is a change in which the desire for love as a central component of life, love both in its erotic and in its humane sense, brotherly and sisterly love among all people, was a powerful motivating force. And it has similarly led to a clearer understanding of the political approaches that are needed to move towards the new goals.

THE TIMES HAVE CHANGED: NEW VISTAS HAVE BEEN OPENED

For our times, one explosive historical set of events may be taken as symbolizing that change, symbolizing both the problems and the possibilities, the aspirations and the difficulties that making our cities as we would have them presents: the events around the spring of 1968. One goal formulation that emerged from those events was "the Right to the City." Seen most broadly, it is the concretization of Herbert Marcuse's realizable utopia in Henri Lefebvre's urban vehicle. The Right to the City meant more than a just sharing of the existing city, an equal right to what is, but rather the creation of a different city, creating in David Harvey's quote from Robert Park, "the city of heart's desire." The call for implementation of a Right to the City makes the vision of a better world into a practical usable political slogan, in the same way that the concept of the Just City is intended to be used. It is a vision that is intended to go beyond redistribution of the existing, into claiming also the right to the fulfillment of the other values that made a humane life worth living, those that the 1968ers wrote on their banners.

The search for the Just City keeps rubbing up against this issue, seeking a goal statement that transcends justice or is uncomfortably defined as part of justice. Susan Fainstein's chapter in this volume addresses this problem in its second half by elucidating the concept of capabilities as well as related concepts of additionally relevant goals.[2] Such concepts go beyond envisioning a Just City as a city in which each individual decision is made in accordance with the principles of distributive justice, however they may be defined. And history has indeed put on the table goals that transcend justice. The events of the 1960s, largely crystallized in 1968, and symbolically represented by the student, worker, and new left actions and philosophies of that year, are manifestations of a historic change in goals. The pursuit of the right to a different city is one enduring formulation representing that change.

The historical importance of that change has been widely noted, although it has been described in equally widely varying terms. The movement is from Fordist to post-Fordist economies, from production relying on heavy manual labor to technologically advanced methods, from nationally bounded to global economic and political structures, from predominantly rural to large new urban settlement patterns, from crude to ever more technologically advanced means of communication and transportation. For our purposes, the critical change was to a society that could technically overcome want and produce enough material goods to satisfy all basic needs for all people, needing in the process a bare minimum of thoughtless or heavy work. It was already a society that was "producing the goods" for the vast majority of its members, if not for all. And yet, when the image of what such a technologically advanced society could produce was held up against existing conditions, growing numbers of people were dissatisfied—not only with the unequal distribution of goods, but also dissatisfied with the human relationships that were involved in their production—the priority given to acquisition and displays of wealth, the narrow focus on economic growth, the concentration of power, the insecurities required to provide motivations for daily work, the destructiveness that underlay irrational expansion, the homogenization producing "one-dimensional" human beings, and the aggressiveness in personal relations and in national practices, particularly leading to war.

In the civil rights movements of the 1960s, as well as the anti-Vietnam War movement, and the student unrest with international repercussions at the University of California Berkeley and Columbia University, all of which were supported by many workers if not their organizations, the disconnect between what was transparently technologically possible and what in fact that technology was being used for, were highlighted. For the 1960s did add a new ingredient to the conventional liberal claims of the earlier centuries. This was made possible by the technological promises of plenty and prosperity that were based on a system in which the exploitation, domination, and racism were concealed but nevertheless central. The one-dimensionality that was the target of the 1968 movement resulted from a system in which profit was derived from never-ending competition and never-ending growth, making possible the fulfillment of actual human needs to an extent that it never before had been, without requiring the on-going creation of more needs to keep the system going. With a new awareness, new claims were expressed in action.

Those claims were not entirely new historically (the Lawrence strikers put "Bread and Roses" on their placards in 1912!) but had formerly been confined to art and culture, as utopian visions of beauty, or as critical visions without political reality. These new claims took seriously the "pursuit of happiness" both as a social goal and as a personal one. They were the foundations of the protests of the 1960s. The earlier claims were included: the right to equality in opposition to discrimination was practiced in personal relations and successfully addressed politically with broad civil rights legislation and political reform. The acceptance of more widespread democratic participation in politics extended liberty, and opposition to domination ended the Vietnam War under the claim for justice and liberty. Separately, these were each major reforms; they came together with anti-colonial revolutions in Asia, Africa, and Latin America after World War II. The Columbia University occupations in New York—involving protests against a University building preempting a public park,[3] an acceptance of the leadership of African-American youth, and an opposition to the use of University resources for research into technologies of death for the U.S. military—symbolized both. The protestors of the 1960s were deeply disturbed at the difficulties of providing an adequate material life, in the face of the reality of increasing inequality in the distribution of goods and services. But they added to that a strong claim to expand the possibilities for the pursuit of happiness, both as individuals and collectively, although the claim was often rather inchoate and expressed more theoretically and philosophically than politically.

On a wider stage, the civil rights movement saw a new historical moment coming and moved towards a comprehensive critique of the system that the 1968ers opposed. Martin Luther King stated it most sharply:

> we have moved into an era where we are called upon to raise certain basic questions about the whole society. We are still called upon to give aid to the beggar who finds himself in misery and agony on life's

> highway. But one day, we must ask the question of whether an edifice which produces beggars must not be restructured and refurbished. That is where we are now.[4]

No political revolutions resulted from the actions of the demonstrators in the streets of New York, Paris, Berlin, or Montgomery and Detroit and Los Angeles. Major reforms, yes; but the idealistic aspirations, particularly of the participants, were abortive. They responded to the one-dimensionality of the world around them, and linked their dissatisfaction to the anti-war and civil rights movements, but not, politically, to the third source of protest, exploitation. Symptomatic was the attitude of the police in the Columbia University building occupations: many treated the students harshly as élitist brats enjoying the luxuries of an expensive education a working policeman could never aspire to. The common nexus which related the students' broad aspirations for freedom and happiness to the more pressing problems of the lacking material opportunities of directly exploited workers, did not come together. In France, workers were directly engaged, but the bulk of the trade unions withheld support from the broader aspirations, refusing to see connections where doing so would have interfered with more pragmatic considerations.

While not a revolution, what 1968 did produce was a new goal, a new demand: the Right to the City. Its pursuit is perhaps the next step in the search not simply for the Just City, but further for the human city, the city to the possibility of which the 1960s opened so many eyes, the city to which it revealed the right might be realistically claimed. As Henri Lefebvre developed the concept, "city" was a synecdoche for society. He stressed—perhaps over-stressed—the importance of urbanization as a major component of change, and saw the possibilities of an urban life as the goal worth striving for, but urban where the cultural contradiction between urban and rural had been overcome (Lefebvre 1996, 2003). Lefebvre writes:

> the right to the city is like a cry and a demand . . . [it] cannot be conceived of as a simple visiting right or as a return to traditional cities. It can only be formulated as a transformed and renewed right to urban life . . . as long as the "urban" [is the] place of encounter, priority of use value, inscription in a space of a time promoted to the rank of a supreme resource among all resources.
>
> (Lefebvre 1996: 158)

Elsewhere, he speaks of the "rights of the citizen as an urban dweller," and says the "right to the city should be complemented by the right to difference and the right to information . . ."[5] Despite his sometimes polymorphous language, it is clear that he is calling for major social change in the city to which the claim of right is addressed, a change away from the "bureaucratic society of controlled consumption," not a bad formulation of what the 1968ers were generally also after.

THE TIMES HAVE CHANGED: OLD VISTAS HAVE BEEN REINFORCED

If the demand for the Right to the City posed a challenge to the established order, that challenge was actively taken up. The 1960s posed a major potential threat to the establishment, even if some today deny it (Marcuse 2008), and the establishment responded, moving in a direction it was in its interests to go in any case. Riots in northern black ghettos were put down by force with use of the National Guard in 1967, Richard Nixon became President in 1969, inaugurating a sharp shift to the right in national policy in the United States. Margaret Thatcher was elected prime minister in Great Britain in 1979, an event often taken as marking the ascendancy of neoliberalism, and the dilution of the welfare state has strengthened since.[6]

URBAN PLANNING, IN RETREAT, NEEDS SUCH NEW VISTAS

So, on the one side, the nature and scope of demands for a change in the existing order grew, while on the other side the forces of the established order strengthened their resistance to change. And the established order has proved strong. One would not guess, from the prevalence of neoliberal policies, that new rights, let alone a Right to the City, are being explored more and more. An examination of urban planning practice in the world today reveals only traces of such an exploration, but the discussion of new goals into which this volume enters, although still much a minority one, and although thus far largely taking place within the academy, is not only an academic one, and has its parallel in the world of social practice. Cities are major hot spots of trouble as well as of promise today, and actions affecting the urban environment can significantly affect what cities are, for whom, and with what consequences. Urban activism and urban planning can play a major role in determining cities' futures. If urban action and planning know where they want to go, what they want to do, and what the possibilities and the constraints are, a real difference can be made in the lives of people. Urban social movements confronting basic social problems seek major changes, but they have declined in strength. Urban planning should be joining with them and helping bring about change.

Yet urban planning is in retreat today. It is in retreat because, as a profession with a vision and a calling to help shape a better future, it is openly under attack from established conservative forces with neoliberal ideologies. It is being asked to confine itself to questions of efficient urban functioning, instrumentalizing social concerns to serve growth and business prosperity. Susan Fainstein's chapter in this volume says it politely: planning has become "modest." It could be formulated more bluntly: formal planning has become opportunistic vision-less, cautious, narrow, small-minded. It has become identified with professional planning, and professional planning has in turn become what professional planners are today allowed to do in the conventional practice of their profession by those in a position to pay them, their employers and their clients, who are overwhelmingly those that hold power in the society.

In this situation, where new vistas have been opened but where the historical balance of forces has turned against them, the contribution of the search for the Just City must be reaffirmed at the same time as it needs to be pushed further. In the face of planning's retreat, Susan Fainstein's call for a return to a normative vision of the city, a vision of what she calls a "Just City," is welcome indeed. She sees planning as a much broader activity than what most professional planners are asked to do, one infused with a set of moral and ethical values, dedicated to the improvement of living conditions for all, to traditions of justice, equity, development of human capabilities: the values she has set forth, and seeks to defend. The point is not whether she has the exactly right formulation for all of these, but whether planning needs to deal with such values, however formulated. As professionals, it is very easy to forget about them, to do the best we can within the limits that are set for us, and not to question those limits. She brings those bigger questions back to the foreground; that is a major contribution today.

While the Just City is an important concept that leads in new and much needed directions for thinking and acting on urban problems, it needs to recognize further goals and further claims. It needs to be expanded to develop and achieve what the Right to the City calls for: the right to a full, free, creative life for all, the right to what the 1968ers so eloquently demanded but fell far short of achieving. But how does this understanding both help to achieve long-term possibilities and at the same time relate to the actions needed to meet immediate and pressing short- and middle-range demands?

A NEW VISTA: THE RIGHT TO THE CITY IN PRACTICE

In developing the content of the Right to the City, practice and theory are on somewhat different if parallel trajectories, with practice outpacing theory.[7] The Right to the City has in fact become a major formulation of progressive demands for social change around the world. Charters for the Right to the City have been debated and agreed upon in various forms at sessions of the Social Forum of the Americas, Quito, July 2004; the World Urban Forum, Barcelona, October 2004; World Social Forum, Pôrto Alegre, January 2005; Revision Barcelona, September 2005.

In the United States, a growing coalition of groups spread out across the country has come together under the name of "The Right to the City Alliance," including groups for the homeless, for immigrant rights, for gay and lesbian rights, and against gentrification. Their inaugural convention, in January 2007 in Miami, has been summarized in an illuminating report in which the relationship between theory and practice has been explicitly addressed.[8] Appearing in force at the United States Social Forum[9] in Atlanta in June 2007, the Alliance had over 250 members present from 20 groups from eight cities, and adopted the following statement of principles:[10]

1. The right to land and housing that is free from market speculation and that serves the interests of community building, sustainable economies, and cultural and political space.
2. The right to permanent public ownership of urban territories for public use.
3. The right of working class communities of color, women, queer, and transgender people to an economy that serves their interests.
4. The right of First Nation indigenous people to their ancestral lands that have historical or spiritual significance, regardless of state borders and urban or rural settings.
5. The right to sustainable and healthy neighborhoods and workplaces, healing, quality health care, and reparations for the legacy of toxic abuses such as brownfields, cancer clusters, and superfund sites.
6. The right to safe neighborhoods and protection from police, Immigration and Naturalization Services (INS)/Immigration Customs Enforcement (ICE), and vigilante repression which has historically targeted communities of color, women, queer, and transgender people.
7. The right of equal access to housing, employment, and public services regardless of race, ethnicity, and immigration status and without the threat of deportation by landlords, ICE, or employers.
8. The right of working-class communities of color to transportation, infrastructure and services that reflect and support their cultural and social integrity.
9. The right of community control and decision making over the planning and governance of the cities where we live and work, with full transparency and accountability, including the right to public information without interrogation.
10. The right of working-class communities of color to economic reciprocity and restoration from all local, national, and transnational institutions that have exploited and/or displaced the local economy.
11. The right to support and build solidarity between cities across national boundaries, without state intervention.
12. The right of rural people to economically healthy and stable communities that are protected from environmental degradation and economic pressures that force migration to urban areas.

In June 2008, the Right to the City groups staged a dramatic "March on the Mayors" at the annual United States Conference of Mayors,[11] including demands ranging from halting gentrification to the rights of day laborers to provision of facilities for gay youth.

The Right to the City slogan has been, if anything, more adopted internationally than in the United States. It is a major campaign slogan for the Habitat International Coalition. Its goals include:

1. Equal opportunity for a productive and freely chosen livelihood.
2. Equal access to economic resources, including the right to inheritance, the ownership of land and other property, credit, natural resources and appropriate technologies.
3. Equal opportunity for personal, spiritual, religious, cultural, and social development.
4. Equal opportunity for participation in public decision-making.
5. Equal rights and obligations with regard to the conservation and use of natural and cultural resources.[12]

The European Urban Charter, adopted in the same tradition, espouses in principle #17, a broad right to "PERSONAL FULFILLMENT: to urban conditions conducive to the achievement of personal well-being and individual social, cultural, moral and spiritual development." In these various formulations we see a mix of old and new demands; many calls are still for "equal" conditions, where "full" attainment might have expanded the claims to take into account the new possibilities. References to "equitable infrastructure of cities" and "distributive equity" are joined to calls for preservation of "cultural memory and dignity," "peaceful coexistence," and a "culturally rich and diversified collective space that pertains to all of its inhabitants." The focus uniformly is however on, "in particular . . . the vulnerable and marginalized groups."[13]

Looking at these statements, eight principles stand out on which the Right to the City movements rests. The first four are clear, the last four emerging:

1. Top priority is given to immediate and basic needs: water, land, housing, work, health care, and to the leadership of people at the grass-roots level in making the decisions that affect meeting those needs.
2. The impact of justice as a goal is clear: words like "equal" and "just" are prevalent. Although in other cases demands are general: words like public ownership for public use, or conservation of natural resources are incorporated.
3. Rights are to be established and implemented democratically and with full participation of all members of society.
4. Rights are seen as inseparable and collective as well as individual. It is of the essence that multiple rights are listed, that the campaigns are not separated but joined.
5. The claims of rights are embedded in an implicit general critique of the existing order, a sense that, putting the claims together, radical changes and transformations in that order are called for, and in fact are needed.
6. The full satisfaction of basic needs is envisaged, without reference to limited resources.
7. The claim for rights to the city, while giving priority to those with immediate and basic needs, is inclusive of broader needs such as dignity and richness of culture.

8. The rights are claimed, not only for those presently excluded from their exercise, but inclusively for all members of society.

These eight principles are major moves forward towards the goals of a better society. Fleshing out the last four, in light of the historical changes that have taken place in the last 50 years, is perhaps the appropriate next step in the search, not only for a Just City, but for a better society of which it must be a part. Some indications of directions in which such a search might go can be suggested.

FROM PRACTICE TO THEORY: SEARCHING FOR THE RIGHT TO THE CITY

The focus of further work in the quest for the Right to the City might well be in the elaboration of the last four of the principles outlined above. These principles were indeed presaged by much earlier work: from the socialist utopians of the early nineteenth century through the work of Marx and Engels and some (but not all) of social democratic and communist theorizing before and after the First World War. As well, many of the liberation struggles against colonialism and racism were concerned with these principles. But they came into sharp focus in the 1960s, when the historical conditions for their fulfillment seemed to emerge dramatically, a situation that produced the theoretical debates around the New Left. In this context as well, Henri Lefebvre popularized the phrase "Right to the City" and its potential meaning,[14] Herbert Marcuse formulated his writings and speeches, and today a number of thinkers among whom David Harvey is probably the best known continue in this vein.

To take the last four principles (5 through 8) listed above one at a time:

> 5. The claims of rights are embedded in an implicit general critique of the existing order, a sense that, putting the claims together, radical changes and transformations, in that order are called for, and in fact are needed.

Here the work is already substantial and represented in the many critiques of neoliberalism in the political sphere and the critical analyses of globalization, financialization, and contemporary capitalism. The need here is to spell out crisply the direct link between the existing order and the conditions that make those claims of rights both necessary and realizable. The necessity may be almost self-evident, and not only to the radical critics, but for those of good will, e.g., many moved by religious beliefs, or holding onto traditional liberal values. The question of realizing these demands brings into play the next principle of the Right to the City.

> 6. The full satisfaction of basic needs is envisaged, without reference to limited resources.

Here existing work is more limited, because it is harder to find the middle ground between the spelling out of immediate ameliorative proposals, such as are usable in actual election campaigns or legislative lobbying, and proposals that sound utterly utopian and irrelevant in the real world. If we use the distinction that Immanuel Wallerstein has been pushing in this context between short-range, middle-range, and long-range proposals (Wallerstein 2008) the need here is to elaborate each: develop concrete immediately actionable demands, spell out the broader programs of which they should be a part, and think through the ultimate possibilities, the concrete utopias, to which an informed and imaginative vision would lead. And the further need is to link these together. To take an example, in order to deal with New Orleans after Katrina, the short-term demands can be for the full and immediate provision of adequate housing for all victims of the hurricane, in the middle range a complete revision of the planning process and municipal decision-making to make it open, democratic, and participatory, and the elaboration of the vision of New Orleans as a model of what a city in the twenty-first century in a country over-flowing with riches could really be.[15]

> 7. The claim for rights to the city, while giving priority to those with immediate and basic needs, is inclusive of broader needs such as dignity and richness of culture.

Here no sharp lines need to be drawn between basic needs and broader needs, between material and cultural needs; indeed the argument is strong that broader and cultural needs are as basic as material ones, although basic material needs have to be satisfied in order for any other needs to be met. Exactly what those broader and cultural needs are need to be further spelled out: what does dignity really mean, what elements of cultural identity are we concerned with, are there limits to sexual freedom, what humane relations among individuals or groups are to be established as socially necessary and what are optional. As well, what do general phrases such as the full development of the human potential, or the expansion of capacities, mean concretely. Again, there is already solid work along these lines, but much more could be done, including work towards comprehensive formulations and, more important, formulations that are usable and understandable and appealing in day-to-day campaigns.

> 8. The rights are claimed, not only for those presently excluded from their exercise, but inclusively for all members of society.

Here we are on little explored territory. Ernesto Laclau speaks of two different systems of life, that of "(1) legitimate citizen, housed and economically productive, deserving of and afforded full legal protection; and (2) . . . a person who is targeted and vulnerable to expulsion and banishment to spaces of exception, . . . not . . . considered deserving of legal protection."[16] Slavoj

Žižek (2008) wrote a short piece in which he spoke of the divide between the Included and the Excluded.[17] It is a useful short-hand for a deep divide, and it echoes an approach the European Union has adopted in seeking "inclusion" as the answer to urban social problems.[18] But these formulations raise major questions on both ends. Is the legal protection afforded the "legitimate citizen" really "full?" Many under-paid, over-worked, or laid off hardly think so. How about productive non-citizens? Are the Excluded totally without legal protection? Many committed lawyers argue the contrary. Are the Excluded in fact excluded from the operations of the system, or are they essential to it, and only excluded from some of its benefits? Are the Included a homogeneous group in which all benefit from the system, or are they in fact composed of many different groupings with quite variable, sometimes even critical, relations to the system? Do not the Excluded have the same broader needs as the Included: in a sensitive article Sarah Dooling quotes the homeless she interviewed as wanting "autonomy, dignity, a sense of community" (Dooling 2007: 46). And to what extent should the goal of inclusion of the Excluded in the system, rather than a change of the system itself—a point strikingly raised by Stokely Carmichael after the major successes of the civil rights movement in the United States, when he asked whether blacks really wanted to be included in the society that has oppressed them for so long, or rather want a different kind of society altogether.

1968 addressed these questions directly, but did not resolve them. When black students asked whites to leave Hamilton Hall in the Columbia building occupations and leave blacks to occupy it and manage it themselves, they were drawing a sharp line between themselves and their counterparts. When Ashwin Desai (2002) argues that "the poors" must take the leadership in the struggle against their oppression, he means it in the traditional sense of the old slogan: "the freeing of the working class can only be the work of the working class." But Desai goes further, as did the black students at Columbia, and sees the effort as not wishing too active participation by non-poors.

The theoretical problems here are difficult. They involve a re-examination of the class structure of society on both the economic and the cultural side, and a questioning of whether the benefits of widespread inclusion are addressed to real or false needs. Does such inclusion address the basic hopes of full development of all faculties or is it shaped by the profit to be derived from manipulating desires for material goods and social status? If the latter, are not the Included themselves victims of the system from which they seem to benefit, and in the long run potential allies of those excluded even from the most basic of those benefits? This is one of the many areas where the theory and the practice of the Right to the City might plausibly come together.

A final area is already widely discussed in theory, but its practical implications need development. Many of the contributions to this book address it through specific cases. The notion of right to the city implies not only a societal change but a spatial change as well, for the ways in which the notion is conceived and justified depends largely on the use of space itself. Interestingly enough, Mustafa Dikeç has the same thought, but reversed in his contribution in this book (p. 83), and therein lies a problem: "the notion of right to the city implies not only a spatial change, but a societal one as well, for the ways in which the notion is conceived and justified depends largely on the very society itself." The relation of space to social justice has been a prime concern of geographers ever since David Harvey's (1973) *Social Justice and the City*; it has likewise occupied many planning theorists and somewhat fewer planning practitioners, and is one of the motivating themes of this book. There is a tendency to make all issues of social justice issues of spatial (or what Harvey calls territorial) justice, which is to limit the concerns much too narrowly. But understanding the right interrelationship between the two is critical in practice as well as in theory, for it has to do with the possibilities and limits on what can be accomplished through actions addressing the use of space in cities. Don Mitchell's (2003) *The Right to the City: Social Justice and the Fight for Public Space*, for instance, makes an eloquent argument about how unjustly public space is used, with particular reference to its use by the homeless, but he comes close to equating social justice with the handling of the spatial aspect of such issues. Resolving spatial issues so as to meet the broad needs of all is indeed a critical element in implementing a right to the city. The tools of urban planning can be a major force for this purpose. Both their potentials and their limits need to be more clearly understood, and further theoretical as well as practical exploration of what is possible and what requires other types of action is needed.

THEORY AND PRACTICE TOGETHER

If work exploring all of the last four principles of the Right to the City and its spatial aspects are needed, the last two, the broadening of needs and the inclusion of all in their fulfillment, open the most immediate questions for practice, for organization and mobilization. Putting the two together, they suggest that, if the needs of the Included can be linked to the needs of the Excluded, a powerful force for change can be brought into existence. The common goal is the satisfaction of all needs, first the most basic needs for survival, then the material needs for an adequate standard of living, and then the broader needs for a full development of the human faculties in a supportive collective environment.

Some are truly excluded from participation in the system; the homeless are perhaps the most extreme example, although even they are often partially included, partially excluded. The lowest paid workers, often in the informal economy, are certainly sufferers from the system: immigrant undocumented laborers in the United States, for instance, are both excluded and included. Most workers are paid less than the value of what they produce; they are included in the system, indeed necessary for it, but also victims of it. Many of the included at higher paying jobs are insecure, overworked, tense, required to forgo satisfaction of their broader needs in the interests of staying included. Many are included economically but excluded culturally; gays and lesbians, may hold prestigious positions but face discrimination in many aspects of their lives. Intellectuals and idealists, arriving at critical positions by virtue of pursuit of their own thought processes or examination of the moral values of their society, are technically included, but support opposition to that in which they are included. All of these have reason to support demands for the Right to the City. The search for how that can come about, what its societal basis is, should perhaps be the priority task in the continuing search for a better society.

Searching for the Just City is an essential part of what needs to be done. Establishing the Right to the City is the logical next step towards its ultimate goal.

NOTES

1 It is tempting, given the argument below, to recall Karl Marx's definition of all history as the history of class struggle, and the present newly-arrived-at stage as one in which the end of class struggle could be seriously envisaged—but that brings us far beyond the scope of this discussion!
2 Although her language does not clearly distinguish it from a distributional approach, i.e., examining which "alternative . . . benefits the least well off."
3 A point that might with hindsight be seen as an affirmation of the Harlem community's right to the city, their city, but the phrase was not by then in widespread use on the western side of the Atlantic.
4 Speech to the leadership of the SCLC, Frogmore, May, 1967.
5 Lefebvre (1996: 170). See also the more extended exposition and commentary in Mustafa Dikeç's, "Justice and the Spatial Imagination," in this volume, who also refers to this language.
6 An excellent discussion of neoliberalism is in Harvey (2005).
7 Mark Rudd, one of the leaders of the 1968 building occupations at Columbia, makes this point: "practice outpaced theory."
8 Available from Tony Samara at tsamara@gmu.edu.
9 https://www.ussf2007.org/en/about. The Forum and its parallels elsewhere is itself a part of the movement incorporating new goals and seeking new forms of organization and coalitions since its founding in 2001 in Pôrto Alegre, Brazil; its call for the 2010 forum in the United States, for instance, includes goals such as: "Work toward greater convergence between working class struggles and progressive movements" (https://www.ussf2007.org/en/vision).
10 Text and brief background at http://www.habitants.org/article/articleview/1988/1/459/
11 See http://www.poweru.org/right-to-the-city-miami-hosts-march-on-the-mayors.htm
12 http://www.hic-net.org/indepth.asp?PID=18.
13 The full text is at http://www.hic-net.org/documents.asp?PID=62.
14 Lefebvre (1996). Although Lefebvre placed great emphasis on the process of urbanization as a turning point, e.g., in *The Urban Revolution*, this issue is not taken up here.
15 I have elaborated this particular example in Marcuse (2007).
16 As summarized by Dooling (2007); see also Laclau (2007).
17 A piece of mine, including comments on his article, is in the following issue (Marcuse 2008).
18 An approach Margit Mayer, among others, has incisively criticized.

REFERENCES

Desai, A. (2002) *We are the Poors: Community Struggles in Post-apartheid Africa*, New York: Monthly Review Press.

Dooling, S. (2007) "Ecological Gentrification: Re-negotiating Justice in the City," *Critical Planning*, 15: 41–60.

Harvey, D. (1973) *Social Justice and the City*, Oxford: Blackwell.

—— (2005) *A Short History of Neoliberalism*, Oxford: Oxford University Press.

Laclau, E. (2007) "Bare life or Social Indeterminacy?" in M. Calarco and S. DeCaroli (eds) *Giorgio Agamben*, Stanford: Stanford University Press: 11–22.

Lefebvre, H. (1996 [1967]) "The Right to the City," *Writings on Cities*, in E. Kofman and E. Lebas (eds) London: Blackwell: 63–184.

—— (2003 [1970]) *The Urban Revolution*, Foreword by Neil Smith, translated by Robert Bononno, Minneapolis: University of Minnesota Press.

Marcuse, H. (1972) *One-Dimensional Man*, Boston: Beacon Press.

Marcuse, P. (2007) "Social Justice in New Orleans: Planning after Katrina," *Progressive Planning*, 172: 8–12.

—— (2008) "In Defense of the 60s: The Pursuit of Happiness is a Goal for All Generations," *In These Times*, 32(8), August.

Mitchell, Don (2003) *The Right to the City: Social Justice and the Fight for Public Space*, New York: Guilford.

Wallerstein, Immanuel (2008) "Remembering Andre Gunder Frank While Thinking About the Future," *Monthly Review*, 60: 2 (June): 50–61.

Zizek, S. (2008) "The Ambiguous Legacy of '68: Forty Years ago, What was Revolutionized, the World or Capitalism?" *In These Times*, 32(7), June.

Index

Page numbers in **bold** represent tables, those in *italics* represent figures.

Lightning Source UK Ltd.
Milton Keynes UK
UKHW020242300722
406604UK00003B/13